PRAISE FOR DREAMLAND

"A brilliant book in which nothing is as it seems, while everything has a rational explanation, and yet, even so, the 'rational' is its own sort of Dracula."
—JOHN LEONARD, *The Nation*

"Nonfiction matter to the novelistic anti-matter of Don DeLillo's recent *Underworld*, *Dreamland* is a brilliantly realized tale of the untold, of U.S. secrecy that's been held like a breath and the farce of its being held too long . . . a must-read for dreamers and skeptics alike."
—*San Francisco Chronicle*

"A rare literary work from the ascendant culture that mingles technology, popular culture, and science fiction with alienation, suspicion, and disconnection from mainstream media, politics, and government."
—JON KATZ, *HotWired*

"This eloquent and frequently astounding book takes readers along on an audacious, circuitous exploration of the desert landscape in and around the most secret military bases in the American West, and of the psychological landscape of fantasy, lore and suspicion that surrounds them. . . . Patton has produced the definitive account of this strange corner of the world and of an even stranger corner of the national psyche."
—HAL ESPEN, *Outside*

"Patton evokes an idealistic covert fraternity whose paranoia and disinformation seeped beyond the borders of Area 51."
—*The New York Times Book Review*

"A psychic probe into the inner nerd of America."
—KEVIN KELLY, author of *New Rules of the New Economy*

"Patton travels beyond the physical location of Area 51 to the psychic location of those who must believe that in the sky exists a world we are not meant to know. . . . A fascinating meditation on delusion and desire, this is an American tale."
— *Kirkus Reviews*

"[Patton] is an observer, a careful listener, a recounter of facts. So he lets UFOs hang there, shadowy forms above the dry bed of Groom Lake, until the closing pages of the book, when he revisits the question and leaves us—refreshingly—with a few open-minded and perspicacious thoughts."
— *The Washington Post*

"[Patton] has written a weird, wonderful, sometimes spooky account of what can only be called a contemporary myth, a 'parable about knowledge and secrecy.' "
— *American Way*

"With one hand on the steering wheel and a pile of brilliantly distilled research on the passenger seat, [Patton] cruises across the arid West and narrates a tale that is curiously epic, frequently humorous, and always entertaining."
— *Tucson Weekly*

DREAMLAND

DREAMLAND

Travels Inside the Secret World
of Roswell and Area 51

Phil Patton

Villard/New York

Grateful acknowledgment is made to the following for permission to reprint
previously published material:

Doubleday, a division of Random House, Inc.: Excerpt from pgs. 166–167 of
Blue Sky Dream: A Memoir of America's Fall from Grace, by David Beers.
Copyright © 1996 by David Beers. Excerpt from Mission with LeMay, by
Curtis E. LeMay and MacKinlay Kantor. Copyright © 1965 by Curtis E. LeMay
and MacKinlay Kantor. Reprinted by permission of Doubleday, a division of
Random House, Inc.

Opryland Music Group: Excerpt from the lyrics to "Great Atomic Power," by
Ira Louvin, Buddy Bain, and Charlie Louvin. Copyright © 1952 by Acuff-Rose
Music, Inc. Copyright renewed 1980 by Acuff-Rose Music, Inc. International
rights secured. All rights reserved. Reprinted by permission of Opryland Music
Group.

Library of Congress Cataloging-in-Publication Data
Patton, Phil.
    Dreamland : travels inside the secret world of Roswell and
Area 51 / Phil Patton.
        p.   cm.
    Includes bibliographical references and index.
    ISBN 0-375-75385-0 (acid-free)
    1. Area 51 (Nev.) 2. Unidentified flying objects. I. Title.
UG634.5.A74P38 1999
001.942'09793'14—dc21   97-48659

Design: Meryl Sussman Levavi/digitext
Random House website address: www.atrandom.com
Printed in the United States of America on acid-free paper

9 8 7 6 5 4 3 2

First Paperback Edition

For my father

Contents

At the Boneyard

"You didn't see that," the officer said.

We were walking amid aircraft in the Arizona desert. It was a bone-yard, like the one in the famous scene in the film *The Best Years of Our Lives*, where planes await the day they will either fly again—perhaps for some Third World air patrol—or be crushed in great machines and melted down into pure aluminum. Hundreds of acres of aircraft shimmered silver in the desert sun south of Tucson—an elephants' graveyard of planes. Military police in blue berets and shiny black boots driving blue pickup trucks patrolled the perimeters. German shepherds rode with them.

The commander of the facility talked too much. It was not a big ca-reer builder, this command, and he talked endlessly about how impor-tant the job they did here was, that it was like a blood bank for aircraft parts, not a graveyard. He hated the word *graveyard*. I suspected he had been given this job because he talked too much.

We walked down the long aisles of Vietnam-era F-105s, their canopies bandaged white like eye-surgery patients, the tiger teeth painted on their noses dulled, the red stars commemorating downed MiGs chipped and peeling. Wherever exposed, the Plexiglas of windows and canopies was scratched, dulled, cataracted. The sun had blistered and flaked the colorful unit symbols, faded the elaborate, delicate green-and-brown mottling of camouflage, and smeared the standard-issue military stencils, NO STEP and RESCUE.

We passed green oxygen tanks stacked in pyramids like cannonballs, ejection seats lined up in a phantom theater, white radomes piled like dinosaur eggs, the black cubes of old altimeters.

There were planes I knew only from putting together models of them in my childhood. Hellcat, Avenger, Hustler, Starfighter, Voodoo, Thunderchief. Aggressive names a kid would like.

In an area they called "the Back Forty" sat acres of B-52s, their backs broken open to reveal green innards. A clown chorus of bulb-nosed helicopters grinned at us as we walked by. Grass and sagebrush had grown knee-high among flattened tires. Birds nested behind ailerons and flaps, jackrabbits lived in jet intakes. Even in broad daylight, the Back Forty is a ghostly place. It's the noise, the creaking of old aluminum, the writhing rustle in the wind of dangling metal and spaghetti wire, the low whistle of an occasional breeze.

I met a man who had worked in the boneyard for thirty years. He was from Waco, Texas, and his skin had been cured to a leathery red-brown by grease and dust and sun. He paused from his work and said, "I always make sure to slap the side of an airplane with a wrench or something to scare out the rattlers and bull snakes and Gila monsters before I get too close."

He was removing an engine. "Some days," he said, "it gets so hot out we have to keep the tools in buckets of cold water just so we can pick them up."

"This whole field used to be covered with '36s," he said—B-36s, the huge bombers that flew over my house when I was a child, growing up during the Cold War, under the aegis of the Strategic Air Command (SAC) and the eagle vision of Curtis E. LeMay. "Had to bring smelters right out here in the field to sweat 'em down. They were too big to move. For days there were columns of black smoke."

Other craft are dragged to the edge of the field, then chopped by guillotine into parts small enough for the smelter, a huge piece of machinery. At its base, the oven emits a liquid as bright as mercury, as thin as water, coursing thinner than you expect metal ever to flow. Molten is too thick and stolid a word for this metal, which quickly cools in ingots that are shipped off to be turned into auto parts, pots and pans, folding lawn chairs.

I spent a whole day at the boneyard. Near the end of it, I caught sight of something in the corner of my vision, a black shape, like a big engine with vestigial wings, with no windows or canopy—no face—no wheels, its shape biological, aquatic perhaps. It seemed greedy and insensate like a deep-ocean-dwelling creature, with the hungry mouth of a ramjet front, as sinister and mysterious as if it had come from another world altogether.

"You didn't see that," the base commander and tour guide said evenly. We paused and looked for a while, then moved on.

I did not know it yet, but I had seen my first piece of Dreamland.

DREAMLAND

1. On the Ridge

Beyond the Jumbled Hills, in the wide Emigrant Valley of southern Nevada, bracketed by the Timpahute and Pahranagat ranges, lies Groom Lake, just one of many dry lakes that dot the desert reaches of Nevada and California, an expanse of white, hard alkaline soil—caliche soil. Rocky Mountain sheep and wild burros often wander onto its surface, and for years the bare weathered horned skull of a sheep sat here, a Western cliché as accent mark. Relentless winds lift small pebbles and drive them across the surface. Once or twice a year, a couple of inches of rain leave a thin liquid layer, a mock lake, shimmering and wavy, whose evaporation rapidly smoothes it to a high polish. The land sat like this for centuries before the asphalt and metal buildings, the wooden barracks and hangars, arrived, turning it into the Shangri-la, the Forbidden Temple of black, or secret, aircraft.

Groom Lake is set inside 4,742 square miles of restricted airspace, and nearly four million acres of bomb range—a space as big as a Benelux nation. It would come to be called by many names: Groom Lake, Watertown, Paradise Ranch, Home Base, Area 51. But the name for the airspace above the lake and the secret test facility and base that would grow there was, irresistibly, "Dreamland." It was this airspace that made it special, the airspace where strange craft appeared and disappeared like whims and suspicions, where speculations like airships glowed and hovered, then zipped off into the distance.

For years it had remained virtually unknown to the public that paid for it, its very existence denied by the government agencies and military contractors that ran it. It was illegal for those who worked inside to speak of it. And fighter pilots flying out of nearby Nellis Air Force Base were forbidden to cross into the Dreamland airspace. They called it "the Box," and if they strayed into it they were interrogated and grounded.

The most famous planes known to have flown at Dreamland were those created by the legendary Lockheed Skunk Works, established by Kelly Johnson. Yet Johnson's successor as the head of the Skunk Works, Ben Rich, told me shortly before his death in 1994, "I can't even say 'Groom Lake.'" To those in the know it was simply "the Ranch," or "the remote location."

A child of the Cold War, growing up fascinated with the mystique of aircraft, I knew the legend already: Here was where the U-2 first flew, and the SR-71 Blackbird and the F-117 Stealth fighter—all in secret. For years only a few grainy pictures of the place—taken surreptitiously from distant ridges or by satellites—served to prove its existence.

On the ridge above Dreamland, I would find I was not alone. Far from it. My fascination was shared by many others—airplane buffs, Skunkers, stealth chasers, Interceptors, like my friend Steve from Texas, like the journalist called the Minister of Words, guys with code names like Trader, Agent X, Zero, Bat, Fox, and others who gathered here, trying to find out about rumored, occasionally sighted, or speculated-upon planes called Aurora, Black Manta, Goldie, and "the mother ship."

Here, too, I encountered the UFO buffs—"the youfers," I would call them. By the late eighties, when a man named Bob Lazar emerged, claiming to have seen and worked on captured flying saucers, Area 51 had become one of the world's best-known UFO shrines.

To some it was the battlefield where the Cold War had been won, an antiwar fought with antiweapons: spy planes like the U-2 that saved us in Cuba in October 1962, or the Blackbird that defused the superpower confrontation in the Mideast in 1973. To one veteran, perhaps cynical, observer of the Pentagon, it was the symbol of a black world run amok, a cult of secrecy grown obsessive, "a secret city," "the last great preserve of cold warriors, a symbol for that wonderful secret world, a testament to how much fun it was to build hugely expensive planes and save the world." To another watcher who was obsessed by all the strange craft in the air, it was a site where "we are flight-testing vehicles that defy description, things so far beyond comprehension as to be really alien to our way of thinking."

To still others, Area 51 implied craft from beyond our planet, recovered in secrecy from desert crash sites or bequeathed in secret treaties with extraterrestrials—craft we were trying to learn how to fly ourselves. For some of the most extreme conspiracists, it was a place controlled by aliens: There had been a shoot-out, the darkest of the stories held, and the aliens who once dined side by side with earthlings in the base cafeteria were now in total control. Or perhaps, a final school argued, it was a place of the grandest deception, a shadowbox of saucer stories playing themselves out in a Punch-and-Judy performance designed to make us accept a final earthly tyranny.

———

Most of the flying saucers or mysterious lights were simply flares, the military argued, used to decoy missiles or illuminate targets at night, and it was plain that some were also landing lights seen through the distance of the rippling desert air. "Yeah, they are unidentified and they fly," one skeptic told me, "and they are sent by a mysterious alien civilization—the Pentagon."

But those watching for secret planes and those watching for alien craft appeared alike in their fascination and their procedures, in their careful accumulation of bits of knowledge, their descriptions of sightings, and, above all, their elusive dreams of a clear view, a clear video image, a clear photograph. "Mystery Aircraft," a 1992 report by the Federation of American Scientists, had observed a striking similarity between the spotters of secret planes and the UFO watchers. The FAS was dedicated to investigating Pentagon waste and excessive military secrecy, but now it had crossed into a new realm of philosophy and cultural analysis to argue that "it is useful to consider mystery aircraft not simply as an engineering product, but also as a sociological and epistemological phenomenon."

———

What had happened to Dreamland was a parable about knowledge and secrecy, about assembling facts and bits of information into a pattern, about learning and speculation. It was about what the Area 51 watcher known as PsychoSpy called "the nature of truth" but was perhaps closer

to the opposite: the absence of certain truth and the abundance of uncertain lore, legend, and just plain "rumint," as the watchers on the
Ridge liked to call it, echoing the military intelligence terms "photint,"
for photographic, "elint" for electronic, and "humint" for human forms
of intelligence. "The signal-to-noise ratio is very low here," one stealth
chaser told me. Or as Steve, the master Interceptor, put it in his Texas
Panhandle locution, "It's awful tough to pick the pepper out of the shit."

It was about mystery engendering fantasy. It was like one of those
empty spaces in the unexplored interiors of continents that medieval car-
tographers had imaginatively supplied with dragons and other monsters.

————

I had driven up from Las Vegas past the F-15s, F-16s, and B-1Bs landing
and taking off at Nellis Air Force Base. A billboard for an upcoming air
show at the base, sponsored by a large casino, promised "An American
Dream Come True." The desert seemed like low-res detail on a flight
simulator game: RISC landscape. This was the country for which God
made cruise control. If you kept your eyes on the horizon, you barely
seemed to move, so slowly did the distant perspective change. You had to
focus on the shoulder, with its blur of sage and silver mileposts, to sense
any progress.

Sometimes on that shoulder, sometimes on the road in front, my
humped cartoon shadow ran ahead and reminded me of the exaggerated
shadows of lunar or Mars landers, taking their own silhouetted pictures
on some distant dry surface. After miles of tilted slabs of stone, striated
like nicely cooked bacon, the only green area was a shock. The
Pahranagat Valley looked like a dark Gothic 1840s vision of heaven, full
of funeral urns and weeping willows, Protestant hymns and early deaths
from typhoid. With its shallow lakes dotted with birds, it offered the
richest land for hundreds of miles around. In the nineteenth century,
horse rustlers used it to fatten animals stolen in Nevada, California, Utah,
and even Arizona.

Past the valley, I came to the little town of Alamo, where someone
wanted to sell a decrepit café, then climbed a long, looping stretch of
road that crested in a high pass called Hancock Summit, where the road
began to descend and the view opened ahead. I caught my breath as suddenly the curtain came up on a vast open westward view across a rising
plain. A dusty white stick appeared pointing straight up in the air. A second later I recognized it as a gravel road, running so straight and so far

and so directly up a slope miles away that in the perspective it seemed like a pole of swirling dust, no longer attached to the land but rising from it like a tightly spun tornado or dust devil.

This, I realized, was Groom Road, the cars sending up contrails of dust as they moved steadily down then up the slung valley, visible mile after mile but barely seeming to make any progress. It was the road that ran up over the Jumbled Hills into Dreamland.

————

We assembled at the trailhead in full view of the DEADLY FORCE AUTHORIZED and PHOTOGRAPHY PROHIBITED signs, beside the motor home that PsychoSpy, the self-appointed watchdog, ombudsman, and tour guide of Area 51, had made his base.

PsychoSpy was Glenn Campbell, author of *The Area 51 Viewer's Guide*, organizer of the Whitesides Defense Committee, publisher of the *Desert Rat* newsletter, the man who had discovered the closest and most accessible viewpoint. He named it "Freedom Ridge" and was delighted when he heard the local guards using that name on their radios. Once you could walk almost up to the base. But after too many curious citizens, including Greenpeace demonstrators protesting at the adjoining nuclear test site, had disturbed their privacy, the Air Force in 1984 went to the Bureau of Land Management, then to Congress, and had large tracts of public land around the base declared part of the Nellis Air Force Base Bomb and Gunnery Range. But two high points, which allowed a glimpse of the base to intrepid hikers, had remained accessible. By the late eighties, the spot began to draw crowds and television crews. That's when the legend began.

Now, in October 1993, the Air Force was applying to take over the viewpoints at Freedom Ridge, and Whitesides Mountain, too. We were heading for Freedom Ridge before it closed for a last chance to look into Dreamland.

————

Hiking up to Freedom Ridge, we dodged the brambly, fragrant sage and the fuzzy, Muppet-like Joshua trees and crossed rocks that seemed inscribed in some alien cuneiform. The perimeter of the base was marked by orange signposts running across the high desert and, on the other side of the barrier, strange-looking silver balls, the size of basketballs, on poles. The lore held that they were motion detectors or other sensors.

Some claim that, thanks to ammonia sensors, these can sniff the difference between a human and a wandering cow or Rocky Mountain sheep. In any case, the exclusion of the public has made Dreamland a de facto wildlife preserve.

I had heard about the sensors and the video cameras and the road sensors, triggered by the weight of a passing vehicle. Helicopters would sweep along the border at sunrise to pick up anyone who had spent the night and sometimes "sandblast" them with downwash from the rotors. I had also heard of the men on the other side of the barrier, in their camouflage uniforms and white Jeep Cherokees, known locally as "camou dudes," who kept an eye on intruders and called in the local sheriff if any crossed the border.

So I kept my eye on the edge, marked with those strange silver balls, until the path rose more steeply and, surprisingly soon, we reached the top. And there it was: I thought of the moment in 20,000 *Leagues Under the Sea* when Nemo's men reach the crest of the atoll and suddenly see dozens of toiling figures down in the circular harbor.

It was all in sharper detail than I had anticipated. The base unfolded beneath us—a line of buildings, fuel tanks, an old bus, the big radar dish, an old bus, and a seven-mile runway—as well as the white horizontal of the dry lake itself. A Jeep came up the road far in the distance, then turned around after a while and left.

I kept finding myself looking back in the other direction, over the valley to the east where the long dirt road puffed up in dust as an occasional car passed. The rooster tails hung in the air a long time. The only black birds we saw were ravens—at sunset eight or a dozen spiraled in formation in the thermals at the edge of the rocks.

———

The ridge at the top was narrow, with a back like a whale, scattered with rocks and tufts of grass and the dead stalks of Joshua trees. It was impossible to pitch tents and hard to find flat places out of the wind wide enough for a sleeping bag. The rocks were black as if a fire had singed them.

I worked to wedge my sleeping bag in between the rocks before dark made it impossible to move. I sorted my gear: My flashlight still bore the masking tape that had identified it at summer camp as my daughter's. My rations were Yuppie MREs (Meals Ready to Eat): Power Bars and trail mix, with a self-indulgent Hershey bar thrown in.

I stood gazing quietly down at the base. "If there are extraterrestri-als," the heavy man beside me said, "it would be the greatest discovery in human history. It would be an intellectual crime not to investigate." He was stern, almost lecturing. He had that chip on his shoulder com-mon to conspiracy buffs. "Please let me finish," he would say too quickly when someone interjected an objection or comment.

We had all agreed not to carry cameras to the Ridge, but now he pulled one out and began snapping pictures. The fine print on the signs also forbade "sketching or drawing" the base, so the notebook and pen I carried would in theory make me equally subject to arrest.

"People get nervous when you mention the idea of extraterrestrials," he went on, while looking through his camera lens. The discovery of life on another planet would shake people's fundamental philosophical and religious assumptions, he said. It would demolish the conceit that we were the be-all and end-all of creation.

But it occurred to me: Rather than how would we explain the exis-tence of other life forms to ourselves, how would we explain ourselves to them? What would they make of us? How could we sum up life here, give a summary of our situation now and the events of the previous half century?

The very possibility of such an encounter, like the prospect of Dreamland stretching out before us, suggested that the exercise of ac-counting for ourselves was a useful one: What exactly would we say to them? Would we explain the atomic bomb and the Cold War, the facing off of two earthly powers and the near destruction of the planet? Or, stranger yet, would we tell of the end of that war, and the deprivation many felt from its lack, the need for an enemy to define ourselves against?

Dreamland seemed an exemplary place to do this. I came to believe that its legend and lore, its language and paradoxes, provided a strange yet appropriate time capsule of a half century of cold war and black se-crecy. Here, the cultures of nuclear power and airpower merged with the folklores of extraterrestrials and earthly conspiracies; their interference patterns formed a moiré of the weird. It was a place from which to see our own planet with the eyes of an outsider.

————

What you called the place revealed what you thought was flying there, and told who you were, just as whether you called a group of islands the

Malvinas or the Falklands, whether you said "West Bank" or "Judea and Samaria" told who you were. People from the Skunk Works called it the Ranch or the remote location. At Nellis Air Force Base, it was the Box, or Red Square. And to hear someone refer to it as Area 51, the name used by the Atomic Energy Commission since the 1950s, meant that his interest was in the saucers. Beneath all these names, the place offered glimpses into the overlapping cultures of UFO lore, of Stealth craft, of nuclear energy and espionage, and into a world whose common ground was secrecy.

———

It was a think tank for Cold War engineering, but with the end of the Cold War—a war that produced its own versions of shell shock and battle fatigue—Dreamland was the center of a great network now in ruins.

Dreamland was the tabloid edge of technology, aptly sited near Las Vegas: It is to technology what Las Vegas is to the everyday economy. It was about playing the long odds. The engineers inside the hangars along Groom Lake were looking for "silver bullets," aiming to strike it rich with superplanes, to hit the jackpot of invincibility. The players were the most important ones in the military-industrial complex: Bechtel, E-Systems, TRW, Hughes, Lockheed, SAIC, and, perhaps the least known of all, EG&G.

Standing on the Ridge picking up lore, I learned about the company called Edgerton, Germeshausen & Grier (EG&G), which did all sorts of things at the test site, from photography to security. The company's founder, Harold Edgerton, was the MIT physicist and photo whiz best known for his stroboscopic photos of bullets passing through apples and milk drops caught in midsplash. He used this technique to photograph atomic explosions for the military and soon his company was providing a variety of services to the Air Force, to the CIA, and then to the Atomic Energy Commission. EG&G has a building at McCarran Airport in Las Vegas from which it operates the so-called Janet Airline of 727s that ferry workers—perhaps a thousand, perhaps two thousand—to and from Groom Lake. And it was EG&G, Bob Lazar claimed, that first interviewed him for a job working on flying saucers.

Another of Dreamland's contractors was Wackenhut, which provided security services—fences and alarms and guards. When Wackenhut was handling security, the guards on the perimeter were called Wackendudes. Then, when it became clear that most of them were from another agency, they became the camou dudes, or just "the dudes."

Among those on the Ridge, the camou dudes grew to near mythical stature. Reports tended to wildly overstate their aggressiveness: Visitors were warned to avoid letting the sun glint off binocular or camera lenses, as if such a flash of light would draw M-16 fire. In fact, the fundamental condition of their jobs, as of those of most rent-a-cops, was tedium. Intruders were irritants and incidents meant paperwork. In the old days, they had ranged freely on public land, working on the principle of deterring the curious before they got near the perimeter. The camou dudes follow, lurk, and watch.

Wackenhut, which also ran the security force at the Nevada Test Site (NTS), had risen to the top of the rent-a-cop business. Like Bechtel or RAND or Mitre, it was one of the specialist organizations that grew up during the Cold War. George Wackenhut, an ambitious former FBI agent, joined with three other ex-agents in 1954 to form the private security agency. He was politically well connected and parlayed his friendships with Florida senator George Smathers, a carousing pal of JFK, and later Governor Claude Kirk, who did most of his carousing alone, into government contracts. An 1893 law, passed in resentment over the use of Pinkerton detectives to break strikes and protests, forbade the federal government to employ private detectives, but Wackenhut's lawyers found a loophole and the company managed to grab contracts for the Titan missile silos and Cape Canaveral. Soon Wackenhut guards were working not just for NASA but guarding embassies around the world, and sometimes handling jobs for the CIA and other agencies that wanted to keep their fingerprints off illicit arms shipments.

I had met Wackenhut men at the NTS and they looked like they spent more time working out than, say, reading. Dressed in temperate-zone camou in the middle of the desert, they did not seem to be students of the natural world around them either. Now that Wackenhut had shifted to an emphasis on more promising business strategies, such as operating prisons under contract for governments eager to "privatize," EG&G found itself in the security business too, supplying guards and even SWAT teams to NASA and DOE facilities.

But the camou dudes at Area 51 seemed to be a mixture of private guards and Air Force guards. PsychoSpy managed to discover—after one of the dudes flashed a Lincoln County deputy sheriff's ID while hassling him—that many of them were deputized by the local sheriff's department. Their notarized deputizations were public record, and he published many of their names.

—————

If by some chance you should see something secret inside the perimeter, the required "oath upon inadvertent exposure" requires you to promise to remain silent, under threat of life imprisonment. Few trespassers were asked to sign it. They were generally charged in county court, fined, and released.

There were constant suggestions on how to get inside the perimeter. They were like a parody of the desperate efforts the CIA and Pentagon made in the 1940s and 1950s to get inside the "denied areas" of the Soviet Union and China, and seemed as wacky as the balloons of Project Mogul, as wild as LeMay's fleet of RB-47s over Vladivostok. The Interceptors figured as jesters in this court of the Cold War: They constantly discussed all kinds of spy schemes, using balloons, rocket gliders, or model car "rovers" with video cameras.

—————

On the Ridge, black-plane buffs, true believing youfers, agnostics and skeptics, radio scanners and heavy optics fans, can mingle in the democracy of curiosity. Even old test pilots would come up here now and then and look at the runway from which they had once taken off. Eventually I became convinced that the two cultures—the stealthies and the youfers—were looking for the same thing.

Standing on the Ridge, I realized that its value grew not out of how much it let you see, but how little: how great the opportunity it created to imagine, fantasize, dream. The irony was that we were spying on spies, peeking in on remote locations at people and machines whose job was to peek in on remote locations. Some of us wore the same camou as the camou dudes, listened in on the same scanners, watched the watchers with the same nightscopes. Such spying was made possible by the Pentagon: by the Internet it had created and the computers its money had developed. In effect, we were self-made spies spying on real spies. And, I would find, others were spying on the spies spying on spies.

Dreamland had been created to devise spy planes to explore Soviet or Chinese missile or nuclear test sites. But in time, Dreamland took on the qualities of the areas it was created to expose—it resembled other, even more remote locations: Kapstan Yar, Tyuratum, Lop Nor. Conceived as a place to facilitate the "penetration of denied areas," it ended up itself behind a perimeter, a denied area.

The perimeter was only the most recent manifestation of our old friend the Frontier: the original settlement line, but also the New Frontier, the Last Frontier. It was the edge of the known, which meant it was the launching point for all sorts of explorations. It was full of the myth and mystery of any inaccessible country.

For a time, I thought of Dreamland as resembling the prints of Hiroshige, such as his *Twenty-four Views of Edo,* in which objects and structures in the foreground seem to get in the way of the views of the landscape. But after looking for a while at this art of "the floating world," as the Japanese call it, one understands that the foreground is also the subject.

Such views inspired Wallace Stevens's haiku-style poem "Thirteen Ways of Looking at a Blackbird." ("There were three blackbirds in a tree / Like a mind of three opinions.") It seemed to me, though, that looking at Dreamland was more like Stevens's jar in Tennessee, "taking dominion everywhere."

"The problem," the Minister would tell me in a phrase I could not forget, "is that the place has no *edges.*"

The light began to fade, the warm sun to soften as it sank. The cold wind flickered around our limbs. We built a privy in a wedge of rocks, draped with a blue tarp for privacy. Two teens spoke in a controlled tone, curious, not fanatic, but credulous, too. They talked of hearing Lazar give a lecture. They spoke of Dulce, in New Mexico, where there were said to be—"were said" was a favorite youfer locution—dozens of aliens living underground. By one account they had massacred their guards and were in control of the complex. There was talk also of the Anthill, an underground installation near Tehachapi, California, and another underground base in that state at Helendale. Someone had been near Helendale recently, and there was talk of things flying in and out of apertures in the concrete at night. "Say there are dozens of aliens underground there too," one teen said, keeping all astonishment, all indications of belief or suspicion from his voice. "Say they're in full control."

As the sun went down, we built a fire, collecting stubs of Joshua trees that looked like soft, oversize pieces of coral but burned with surprising surges of flame and then a fitful glow.

The UFO types talked about Lazar. A fat girl talked about wanting to see strange things—not UFOs themselves, but weird UFO types.

As night fell, the lights came on in the base below, where personnel were probably watching television, amid the inevitable military tedium that attends even the most exotic of projects, more intently than they watched the few people, high above Dreamland, watching them.

The campsites were scattered, miscellaneous, like the social dynamic. In the middle of the night, I woke to hear the fat girl groping her way down the ridge, sleepless and grumbling.

I found myself drawn back again and again to the perimeter. One of the odd effects of visiting the Ridge was that it seemed to make visitors feel an impulse to investigate, in a vigilante way. So I came to fantasize: I saw myself as a mock version of one of those explorers charged by Congress and the Army Corps of Engineers to chart and record distant reaches of the West, men like John Wesley Powell or Clarence King. The idea of a travel account of a place you couldn't physically visit was irresistible. But I got a surprise: The place seemed to spin me away from me the more I found out about it. And I became more fascinated with the watchers than with watching.

There is an Indian petroglyph, a spiral, found on ancient rocks in the Nevada desert, that is thought to represent language. This would be my spiral: out and then in.

2. The Black Mailbox

The next morning we paid the obligatory visit to the Black Mailbox. It is found near milepost 29.5 of Highway 375, about twenty-five feet west of the pavement: a large round-topped mailbox painted black.

The box belonged to rancher Steve Medlin, whose cattle had the right to cross into Dreamland—and did. They also lurked by the side of the highway, looking positively eager to be mutilated by aliens, and loped across the road to endanger rental cars driven by UFO tourists. "Stealth steers," someone called them.

Bob Lazar used this mailbox as a convenient landmark to direct viewers to watch for the appearance of the craft he said flew from Dreamland. But from a mere landmark, the Black Mailbox quickly became an icon as hundreds of watchers flocked to the area hoping to catch a glimpse of saucers rising above the mountains from the Groom Lake base.

Could any symbol be handier than a mailbox? This large, classic curved-steel container, fastened with a small silver-colored padlock and puckered by the passage of .22 caliber projectiles, was a key trope of the information age, a repository for missives—some official, some personal, some commercial, the believable and the exaggerated. It was the perfect symbol of Dreamland. Of the whole black world, the black budget! It suggested blackmail and Men in Black and black helicopters.

What you could see from the Black Mailbox "was better than having sex for the first time," Gene Huff would say. "Not for the second time, but the first."

Huff recalled that when Bob Lazar took him out there the light from the mysterious object over the Jumbled Hills was so sudden and bright they instinctively moved behind the open trunk of the car, to shield themselves. The next Wednesday night, they returned and the disc staged an even more breathtaking performance, blinking and with each blink seeming to jump toward them.

And this was the remarkable thing, the linchpin of Lazar's credibility: He could tell when the saucers would come. After the reports began to appear, more and more people made the pilgrimage to the Black Mailbox to enjoy the same thrill.

They saw everything from red darters to orange orbs and green glowing discs that hovered, turned suddenly, and shot away at incredible speed. They thrilled to "objects that glow with an amber light and flitted like fireflies," to dots that "performed zigzag movement incomprehensible in terms of conventional aerodynamics." They talked of HPACs—"human-piloted alien craft"—captured saucers, flown by humans.

Their accounts appeared on the Internet, full of a sense of menace—more in anticipation of the camou dudes than from their actual behavior. They all had the same sense of being among the first ever to see. And as a rule the farther they had traveled to get there, the more they saw.

Was it just because they had come all the way from Norway that one group worked itself into a lather of twenty or thirty sightings in one night, none of which was recorded on their videotapes? The group posted on the Internet a verbose account of their visit to the Mailbox. The account tells of dozens of saucers and a sky filled with "lazars." I nearly jumped when I noticed this usage—was this a spell checker error or simply imperfect English? The "lazars" crisscrossed and danced across the sky in all colors—laser beams. They thought they saw fake clouds, generated by machinery, which hid the saucers from view, lenticular clouds produced by weather manipulation technology.[1]

> . . . the first sightings were lights bouncing . . . We watched flickers, flashes, and sparks—also tremendous rapid "streaks" of light from base to cloud. . . . Now for the following time until 2:30 A.M. we were having continual sightings. Up to six ships at a time, appeared . . . We grew accustomed to the ships in such a short minimal period of time. After an hour and a half, we were "used" to them . . . There was no "threat"

no nothing, just playful, curious encounters—goes to show HOW fast we humans can grow accustomed to things. . . .

I turn and there's a beautiful "green" ship hovering . . . Then another one came in to the right, an orange one . . . [then] a green object, more Pleiadian shaped than the others we had seen . . . At the bottom of it, two bright lights in motion, but connected to the whole ship, all in orange colors . . . It vibrated and was as if it were alive . . . there was definitely a feeling of life & intelligence . . . for several minutes we were paralyzed in joy and disbelief . . . This particular sighting was the longest one of them all, and gave us real time to "tune in" to it, and become "acquainted" with its presence . . . We were feeling so relaxed about it all, didn't seem at all strange that we were there with UFO's, and the next thing we thought was like: ". . . so, now what?"

———

The visiting youfers felt this was their landscape. To Sean David Morton, self-proclaimed "UFO authority" and erstwhile astrologer, predictor of earthquakes, and channeler from Hermosa Beach, California, it was the only place in the world where you could see flying saucers on a regular schedule and therefore the only place to which he could lead people for good money. Another buff, Gary Schultz, proclaimed himself "the world's authority on Area 51." He established a business guiding tourists on what he called "Secret Saucer Base Expeditions" and had the temerity to rename one of the mountains nearby after his girlfriend Pearl. No one else called Whitesides "Pearl's Peak."

Yet the citizens of Rachel, Nevada, in Lincoln County rarely saw anything at all out of the ordinary. Nor did I ever see anything that appeared not to be a flare or a helicopter or other distinct craft. Of course the ordinary included all the craft flying from Nellis, dozens of planes from the base's Red Flag, Green Flag, and other exercises roaring over the area east of the forbidden Box.

The Interceptors wondered, Wasn't it odd that the schedule of saucer sightings corresponded so closely to the schedule of flights from McCarran to Groom Lake? Didn't "Old Faithful," the UFO that appeared each Thursday morning so predictably for Morton's customers, coincide with the schedule of the early Janet Airline flight from Las Vegas? Didn't a lot of the green lights suggest magnesium flares dropped by fighters to decoy heat-seeking missiles or illuminate ground targets?

For the black-plane buffs, the sightings tended to be more widely dispersed, from as far away as Beale Air Force Base and Mojave in

California. Supersonic planes, after all, could take the width of a good-size western state just to make a turn. The craft seen near Edwards Air Force Base in California would soon be in Nevada. Some saw hovering wings in Nevada near Pahrump, others around Goldfield. And Agent X spotted a bat-winged airplane over the town of Alamo, just up the road from Rachel. Dreamland was simply the end of a corridor that ran back to California's aerospace center in the Antelope Valley, to Edwards and Palmdale's factories, to which the contractors had moved from their original urban factories in Long Beach, Culver City, Burbank, and Santa Monica.

For a time, *Aviation Week* would report in great detail such sightings under headlines like POSSIBLE BLACK AIRCRAFT SEEN FLYING IN FORMATION WITH F-117S, KC 135S, and the details would make hearts beat faster. Some of the hearts were in the medal-encrusted chests of Air Force generals, who expressed displeasure to the editors. In any case, when correspondent Bill Scott was shifted from southern California to the magazine's Washington bureau, such articles became fewer and notably less speculative.

Some sources for the stories described distinctive contrails—the "doughnuts on a rope" said to be characteristic of the new high-tech "pulse detonation engine," which was a real enough technology but of unclear technical maturity.

There were sounds as well as sights: the "Aurora roar" or the "pulser sound"; "a sound like the sky ripping"; "a very, very low rumble, like air rushing through a big tube."

The black-plane watchers and the youfers often stood side by side, looking at the same sky, seeing different things yet uttering a common cry: "Did you see that?"

———

The sign that warned NEXT GAS, 110 MILES was a good enough reason to stop in Rachel, up the road from the Black Mailbox. But Joe Travis and his wife, Pat, who had taken over the Rachel Bar and Grill in 1989, didn't sell gas. They cleverly renamed the place the Little A"Le"Inn and packed it with pictures of planes and UFOs, patches of military units, saucer paintings, UFO models, and such knickknacks. They had a "stealth bomber" patch that showed—nothing. There were painted portraits of aliens by Jan Michalski, an armless Belgian who lived in Nevada. On a small shelf they established a lending library of UFO- and stealth-related books and videos. Behind the Inn stood trailers with rooms to rent, done in a style that could be called generic crime scene. The only thing missing was the chalk outline on the floor.

Most days, Chuck Clark was there. "Chuckie"—as the Interceptors derisively called him—saw his first UFO in August 1957, near his home, just six miles from the Skunk Works in Burbank. There was a flock of them, he told me, and he recounts how F-89s were scrambled to chase the shapes. A crowd had gathered to watch.

He came to Rachel to pursue his study of astronomy in the clear air, and his interest in secret airplanes and flying saucers was just a sideline. He had seen Aurora, he said, one cold winter night, and he talked of how the aliens might come from "another dimensional reality" or how they might be time travelers. He was calm about these possibilities, as if including them in his analyses just to be fair.

It was unfortunate, however, that when Clark grinned he turned into Howdy Doody, a grin he must have had as child, fine on a freckled boy of six but disturbing on a man of fifty, and suggesting—it wasn't a charitable thought but it was an inevitable one—an arrested development. This, I suspect, is why the diminutive "Chuckie" managed to stick.

————

According to the map in the phone book, Rachel was compounded of triangles, although its street plan was not readily discernible from the first view of the trailers beside the desert, like a cove full of boats. One side of a triangle was Groom Road, the back entrance to the base. The Little A"Le"Inn anchored the north side, the rival Quik Pik the south. And at the center of the town stands a radiation recording station that measured possible fallout from the nuclear test site to the southwest, set neatly upon a little plot the way the statue of a Confederate soldier might be placed in a small (but never this small) town in South Carolina.

Civic spirit in Rachel is aptly represented by the most popular contest at the annual town fair. A checkerboard is marked off in the dust with numbered squares, and after bets have been taken, chickens are released. The object is to correctly name the square on which a chicken will first excrete.

Once the town was on its way to "site." That is the Nevada map euphemism for ruin. (Ghost town generally indicates a "site" brought up to tourist ruin standards.) Then in 1973 Union Carbide began mining tungsten and the town, once called Sand Springs, was reconstituted, like a dried shrimp in a science kit, then renamed after the first child born under its new economy. But young Rachel Jones would die just three years later, after her family had moved on—a victim of the Mount St. Helens eruption. Place of death was recorded as Moses Lake, Washington, the site of another secret test area, used by Boeing.

The Inn was renamed after the Lazar craze began to bring UFO tourists to the town. That was Joe and Pat's initial marketing inspiration. The rest flowed from that: the coy "Earthlings Welcome" greeting, the collections of alien masks and UFO snapshots, the menu with "Alien Burgers." Joe let it be known that he had once worked at the base, and did not discourage the impression that the Inn was the prime watering hole for workers at Groom Lake. And Pat told eager tourists and press— the *Weekly World News* and later *The Wall Street Journal*—that she believed the place was guarded by an alien named Archibald. Behind the bar where Joe Travis always stood, beside the sign that reads THANK YOU FOR HOLDING YOUR BREATH WHILE I SMOKE, was another message to visitors: WE DON'T HAVE A TOWN IDIOT. WE ALL TAKE TURNS.

In February 1993, Joe and Pat decided to hold a conclave of UFO buffs, which they boldly titled "The Ultimate UFO Conference." Bob Lazar arrived with a female companion, in a Corvette, and Gary Schultz spoke. Norio Hayakawa, creator of the *Secrets of Dreamland* videotape, played country-and-western music in a corner. It was cold and windy, but the crowd outgrew the Inn. Joe set up a large tent outside and when he was asked where he had gotten it he said, "The boys at the base lent it to me."

———

At the other end of town was PsychoSpy's trailer. Glenn Campbell, aka "the Desert Rat," had been a computer programmer for a successful software company on Boston's Route 128 when, in January 1993, fascinated by Lazar's story, he moved to Rachel. The anagrammatic quality of the nickname, psy and spy, struck me as right on: This guy was different from most of the on-line characters swapping lore.

Glenn was in his activist mode that week, decrying secrecy and waste. His circulars opposing the takeover of Whitesides and Freedom Ridge proclaimed the base "a sacred temple to waste, inefficiency, incompetence, mismanagement, and maybe even fraud." It was absurd to pretend that a huge base didn't exist, he argued, when in fact anyone with breath enough to make it up the mountain could see it. You can't say about a whole base "You didn't see that" and have credibility. The government's policy of denial was breeding mistrust; the government was alienating its own citizenry. "The stories of alien spacecraft at Area 51 cannot help but thrive," Campbell argued.

————

Driving back down to Las Vegas I passed through rain and saw a double rainbow off to the east, arched from mountain to mountain. I wouldn't have believed it had I not seen it myself, as the sighting reports say.

At home much later, when I listened to my tape recording from my time on the Ridge, what came through was the noise of the wind, hissing, flickering, licking. Much noise, little signal. Or was the noise itself the signal?

3. "They're Here!"

In 1989, what seemed a clear signal emerged at last from the noise around Dreamland. Bob Lazar claimed to have worked on flying saucers hidden near Groom Lake. The gawky technician's story grabbed the attention of not just wide-eyed saucer buffs but a wider audience of the curious. Some believed he was telling the absolute truth; others were intrigued by the belief that he *could* be telling the truth. Bob Lazar brought to the borders of Dreamland people who had never heard of the Skunk Works.

In person, or on radio or television, the unassuming Lazar broadcast a believability that grew from his lack of stridency. Calm, almost diffident, he worked a charm that fascinated even those it did not convince. Tom Mahood, a hardly credulous engineer, who researched many of Lazar's claims and found holes in the story of his life, never lost the sense of how subliminally persuasive the man was. His matter-of-factness lent possibility to a story that rendered in cold print seemed outlandish and weird.

In essence, that story went like this:

I saw flying saucers in Dreamland. I worked on flying saucers owned by our government in an area called S-4, at Papoose Lake, south of Groom Lake. I thought I was going to work at Area 51 but was taken in a bus with blacked-out windows to a place where I saw the saucers.

I learned of antimatter reactors used to bend gravity waves fueled by element 115, a reddish orange substance, of which we have about

500 pounds and which comes in discs the shape of half dollars. I had one but the government stole it back.

I saw golf balls bounced off the gravity wave the reactor from the saucer generated. I was allowed to read strange documents—autopsy images of aliens, and a history of the earth as viewed from Zeta Reticuli where the aliens came from.

I saw my fellow workers wearing security badges with one light blue diagonal stripe and one dark blue and the letters MJ. My supervisor had one that read "Majestic."

I saw little chairs in the saucers that suggested little creatures—aliens.

Once I walked by hangars and caught glimpses of—I think—a little alien. But I'm not sure. "It could have been a million things," [the supervisor] said. But I think I saw one.

It began with a chance encounter with Edward Teller, the father of the H-bomb and godfather of Star Wars. Lazar had been working in Los Alamos, New Mexico, for a contractor to the physics labs there called Kirk-Mayer. His job involved particle detection equipment—Geiger counter stuff—and was linked to the Meson or Positive Proton Lab. Locals remembered him as intelligent, kind but a bit of a con man, trying to rustle tools and funds for another project.

In his spare time Lazar had designed a "jet car," a weird mating of a Honda CRX and a jet engine. The local paper, the *Los Alamos Monitor*, had done a story about Lazar and his car, right there on the front page, and on June 23, 1982, the day after the story appeared, Lazar went to a lecture Teller was giving in town. Before the lecture, he spotted Teller reading the *Monitor*. "That's me you're reading about," Lazar told him, and chatted him up.

Several years later, after his marriage had dissolved and his finances gone to rack and ruin, after he had been let go by the contractor in Los Alamos for using government equipment to work on the jet car, Lazar wrote to Teller seeking work. He had moved to Las Vegas in April 1986, in an attempt to start again. On April 19, he married a woman named Tracy Anne Murk at the We've Only Just Begun wedding chapel of the Imperial Hotel. Two days later, his first wife committed suicide, inhaling carbon monoxide in their garage. In October he declared bankruptcy. With the bankruptcy and new marriage, Lazar had begun to put the past behind him, to repair his life and his self-image. Teller would direct him to the people who hired him as what Agent X would later call "the Mr. Goodwrench of flying saucers."

Teller called, saying he didn't have any jobs for physicists but knew someone who might. Fifteen minutes later Lazar's phone rang again. It was someone from EG&G, inviting him for an interview that led to the job at S-4.

Lazar would brag that at the interview he had "dazzled" them. Who were they? EG&G hired him, but his ultimate employer, he said, was listed as the Office of Naval Intelligence. Lazar was able to produce a W-2 form bearing a payer ID number assigned to the Navy; it recorded an annual earning of $977.11.

In December 1988, Lazar said, he began work at S-4, which was ten or twelve miles from Area 51.

———

Bob Lazar liked to feature himself as physicist, and in his most widely circulated photograph he presented himself, chalk in hand, in front of a blackboard covered with abstruse equations, like Oppenheimer or Teller. He claimed attendance at MIT and CalTech and said he had two master's degrees. He talked of "getting back into physics," as if he had been a major lab scientist, and referred to Edward Teller as "Ed." But he was not a physicist in any professional sense. He had made his living as a technician and later as the owner and operator of a fast-photo processing outlet.

———

Gene Huff first knew him as "Bob, the photo guy." Huff was a real estate appraiser in Las Vegas who like many in this business used Lazar's photo shop to develop pictures of houses. Usually, Lazar's wife, Tracy, delivered the photos, but sometimes Lazar would show up himself. On these occasions Gene Huff and Lazar would talk. They were both interested in explosives and were part of a group that occasionally went into the desert to set off big explosions. Huff once saw Lazar mix up some nitroglycerin at his kitchen table.

Lazar liked fast cars even better than big booms. He once drove a 1978 Trans Am powered by hydrogen, and he built the jet car that had been featured in the *Los Alamos Monitor*, a Honda CRX with a jet engine in the back and the license plate JETUBET. He borrowed two thousand dollars to build a jet-powered dragster, a thirty-two-foot-long conglomeration of steel pipe with a surplus Westinghouse J-34 jet engine from a Navy Banshee fighter. It could run at over four hundred miles per hour.

———

Tom Mahood, the most relentless archivist among the Interceptors, traveled to Los Alamos and Las Vegas to document Lazar's life. He learned that Lazar had been born in Coral Gables, Florida, then adopted. No records exist proving he attended CalTech or MIT as he claimed. He had attended Pierce Community College in California and had a mail-order degree from a place called Pacifica University. But the *Los Alamos Monitor* did report that he was a physicist at the Los Alamos Meson Physics Facility or, as he called it, the Polarized Proton section.

A geeky-looking character with large glasses—your classic nerd— Lazar claimed that at S-4 he was assigned to figure out the propulsion system of flying saucers. There were nine different kinds of saucers, he reported, and he gave them nicknames like "the sport model" (a term taken from a Frisbee brand name), "the Jell-O mold," and "the top hat."

The saucers traveled by means of a gravity-wave generator, involving a reactor of some sort, and an amplifier that directed the waves. Lazar took credit for identifying the fuel on which the reactor ran as "element 115," a heavy rust-colored substance with an atomic weight far greater than that of lead. He had surreptitiously pocketed some of the supply of element 115. It was to be his ace in the hole, his way of proving his story, but it had been stolen from his house.

Yet Lazar—and here he dons his role as "physicist"—expressed shock at the crude state of the research at S-4 and the low qualifications of those doing it. They tried to make a saucer run on plutonium instead of element 115, he had heard, and the result had been a disaster. And they had foolishly cut open a reactor while it was operating. It was the resulting deaths, in 1987, that had opened up a job slot for him. Lazar declared that, in only a few days, dealing with "materials that were— pardon the pun—totally alien," he had figured out the operating principle of the saucer's antigravity reactor. He was a big-time physicist at last, working on a project even bigger than Edward Teller's.

———

If anything lent credence to Lazar's story, it was that he knew when the flight tests for the saucers were scheduled—Wednesday nights, he reported—and they would appear over the Jumbled Hills between the Groom Lake road and S-4.

He took Huff, John Lear, and others up to see the saucers fly. On March 15, 1989, using Lear's RV, Lazar, Huff, and Lazar's wife and sister drove up to Groom, turned out the headlights, and headed down the long sloping dirt road that runs up into the mountains and to the Groom Lake perimeter. Looking through a telescope Lazar soon reported an elliptical light rising above the mountains between them and S-4. The light began jumping and dancing around, then came to a dead stop and hovered. But after just a few minutes, the light slowly sank back down behind the mountains.

Huff and Lazar returned a week later. Huff recalls that Lazar's wife, Tracy, and a friend named Jim Tagliani joined them. The next Wednesday, the group rented a Lincoln Town Car and returned to the area. "We turned our lights off, and went in about five miles on the Groom Lake road. We pulled off on a side road and unloaded our video camera, telescope, binoculars, et cetera, out of the trunk, and we left the trunk lid open."

The disc came up around the same place, and this time it staged a breathtaking performance. It repeated moves similar to the week before, but now it came down the mountain range toward them. At first it seemed far away, then they'd blink and it would seem a lot closer, then blink again and it would seem even closer. There was no sense of continuous movement; the disc simply "jumped."

The object was also incredibly bright, so bright that Huff remembered how they moved behind the open trunk of the car, reflexively seeking protection as if from an explosion.

Lazar told them this motion was due to the method of propulsion and the way it distorts space-time and light. He also explained that the bright glow of the disc was due to the way it was energized. "An explosion was the only thing, other than the sun, that we had ever seen be that bright," Huff recalled. They took a videotape, and the camera recorded the sighting at around eight-thirty. Eventually it set down behind the mountains, and they left. Huff had never seen anything like this in the sky in central Nevada.

The next Wednesday, Lazar, Tracy Lazar and her sister, Huff, and Lear arrived shortly before dusk. Huff recorded, "Numerous security vehicles were sweeping the roads that the cattle ranchers use to round up their cattle after open-range grazing. It seemed that this night, more than the previous Wednesday nights, they wanted to make sure no one was out-side of Area 51.

"We tried to sneak in using our usual 'stealth' mode," Huff went on, "but security saw our brake lights and began to chase us. We tried to beat them out to the highway, but they came from all directions and ulti-mately we had to stop. We told them we were simply out there stargaz-ing, which they didn't believe for one moment. They agreed that they couldn't chase us off of public land, but simply said they would 'prefer' that we retreat back up to the highway. They issued us a copy of a writ-ten warning that said we were approaching a military installation and cited the relevant statutes, including the penalties for taking pictures of the base."

The group returned to the paved highway, but a short time later a Lincoln County deputy named LaMoreaux pulled them over and asked for identification. He took their IDs and radioed the security base station. It was obvious, Huff felt, that the guards and the sheriff's office worked together. But the deputy finally let them go.

The next day, Lazar got a phone call. His supervisor at S-4, Dennis Mariani, had learned of his latest expedition to the Black Mailbox. He was to report for debriefing. "When we told you this was secret," Lazar recalled Mariani saying acidly, "we didn't mean you should bring your family and friends to watch."

Mariani drove Lazar the forty-odd miles from Las Vegas to the de-briefing at the old Indian Springs airfield. Lazar was told the test sched-uled for the night he was caught had been canceled. But according to Lazar, they neither fired him nor revoked his security clearance. He sim-ply never went back to work.

Lazar had violated all the security rules, yet had not really been pun-ished. He had been warned about security from the beginning of his employment at S-4, he said, by men holding a gun to his head. He was sure all along he was being watched. Security people visited his house again and again and dropped in on his friends. The disc of element 115 he had secreted disappeared from his home. When Lazar talked with Gene Huff at Huff's home, both men felt sure they were being overheard by listening devices, so instead of speaking aloud they passed notes which Huff burned afterward. In the notes, they referred to each other as Bufon and Gufon, a joking reference to the UFO organization MUFON (Mutual UFO Network).

But Lazar believed that his wife was having an affair and that it was this, and not any of his security breaches, that led to his termination. He thought that the security forces at S-4 had recorded and transcribed his wife's phone calls, and that in their judgment, he was probably unstable and a potential security leak. By May, the couple would be separated.

At that point, Lazar decided to go public. He had already recorded a video interview with newsman George Knapp, but it was for "safekeeping," and never aired. In May, Lazar agreed to another interview, this time for broadcast, but in disguise.

Not until November 10, 1989, when Lazar appeared under his own name and showed his face, did the story have a major impact. On November 21, Knapp and Lazar together appeared on *The Billy Goodman Happening*, an AM radio show with a huge audience. On November 25, KLAS-TV ran a two-hour compilation of the Lazar interviews and other clips under the title *UFOs: The Best Evidence*. On December 20 he was back on the Goodman show. By then the story was getting international coverage.

Lazar increasingly relied on Huff as his confidant and handler in dealings with the press. He wanted someone else to get a confirming look at Dennis Mariani, his supervisor, so he set up a meeting with him at a Las Vegas casino, and without telling Mariani, he brought Gene Huff along.

Huff had been told to look for a bulky-bodied ex-Marine type with a little blond mustache. Huff found Mariani sitting at the blackjack table between two large-breasted women and behaving oddly: He was not looking at them at all. In Las Vegas this seemed highly aberrant behavior. But even worse, Mariani pretended not to recognize Lazar, and the meeting never came off. Perhaps Mariani had noticed Huff; Huff had caught sight of another man with him who he said looked like a security agent.

―――――

Early the next year, Norio Hayakawa, a UFO researcher who had seen the KLAS-TV broadcasts, brought Lazar to the attention of Nippon TV, and in February 1990 he took a Japanese crew to Las Vegas. They interviewed Lazar at what was described as his house, but Hayakawa thought it felt strange. There wasn't much furniture, and the place didn't look lived in. A man introduced only as "a friend" sat beside Lazar and even followed him to the bathroom. The man wore some kind of beeper on his belt.

Lazar suggested a time and place the crew could watch the saucers fly, film them, and confirm his story. He sent them to the Black Mailbox. At 6:45 one morning they saw a bright light over the Groom Mountains. At 8:15 a brilliant orange orb jumped erratically.

Lazar agreed to appear live on Japanese television and had even accepted plane tickets to Tokyo for himself and Gene Huff. But Hayakawa waited for him in vain in the terminal at LAX. Lazar never showed. When Hayakawa telephoned, Lazar told him that he could not come; his life was in danger. His tire had been shot out when he was on the way to the airport. Significant money had changed hands as well as the plane tickets, and to save face, the network set up a telephone link so Lazar could at least answer phone-in questions live during the show. Some thirty million Japanese viewers saw the program.

———

Then something even weirder happened. In April 1990, not long after the Japanese show, Lazar was arrested in Las Vegas for pandering, an obscure charge akin to living off immoral earnings. He was convicted on June 18.

Lazar had long boasted about a legal brothel he had wanted to start when he was still in Los Alamos. He planned to call it the Honeysuckle Ranch, and there is some evidence he filed the legal papers necessary and even had T-shirts made up bearing the name. But whether the brothel idea was simply a running joke, a fantasy, or a half-realized business effort remains unclear.

The Las Vegas episode had begun when, after his separation from Tracy, Lazar, in Huff's singular phrase, "took comfort with a hooker." He became friendly with the girls and, according to the charges, ended up working with a prostitute named Toni Bulloch and helped set up a computer database for a brothel in the Newport Cove Apartments, a Spanish-style complex near the airport. Sentenced to community service, Lazar helped install computer systems for worthy organizations and showed up at a Las Vegas children's museum to give courses in computing.

Those researching his probation report found that all government records about Lazar's past had been sealed away under a federal "need-to-know" restriction, further intriguing the believers. Was it part of a plot to silence Lazar, make him disappear? Had he been set up for the whole charge? Or was the government just protecting its own?

———

This mystery, possessing the part mirror, part pewter surface of Lazar's Sport Model itself, made his story intriguing. His manner had the same effect: a combination of bright highlights and dull spots. To John Andrews, the veteran Interceptor, Lazar's appeal lay in the fact that he was one of the rare UFO witnesses to say "I don't know" about parts of his story. While most UFO stories were dogmatic in their detail, Lazar's was full of gaps and limits. He refused to speculate on the source of the saucers, for instance.

There were problems with his story, of course. As Mahood had shown, his CV did not jibe with reality. The Social Security number on the W-2 form did not belong to a man named Robert Lazar.

To those familiar with military programs, the descriptions of the saucer program Lazar gave in his interviews included elements that seemed unlikely. He was shown more than was believable, they thought. Special access programs were famously "compartmentalized." The engine people were not allowed to see what the wing people were doing, and so on. At Groom Lake, for instance, the SR-71 ground crews never knew the destination of the plane. But for some reason Lazar was offered glimpses of many different aspects of the program. Sometimes he said he thought he was allowed these as tests of his loyalty.

After he went public, Lazar took two lie-detector tests, but both were inconclusive. At best, the tester said, Lazar believed what he was saying, but he might have been relaying on information given to him by someone else.

Tom Mahood's researches into Lazar's background had revealed the deception. However much power "they" had to erase his past, it is inconceivable that they could have removed Lazar from all copies of MIT or CalTech yearbooks and directories.

———

Yet even with all the problems, Lazar's tale drew an increasing audience; he created a fascination even among skeptics.

In his essay "Lazar as Fictional Character," PsychoSpy got to the core of Lazar's appeal: that willingness to admit the limits of his knowledge, the restraint in his speculation, and the almost eerie consistency of his tale through interviews over the years. He was perhaps like a witness who tells too good a story in court. Yes, there were a few places that didn't gel. Once Lazar said that one of the saucers "looked like it was hit with some sort of a projectile. It had a large hole in the bottom and a large hole in the top with the metal bent out like some sort of, you know, large-caliber four- or five-inch [shell] had gone through it." But in most interviews he said, "None of the discs looked damaged to me."

Still, it was remarkable how consistent Lazar was in his telling, and PsychoSpy praised the "impressive coherence and integrity of the story itself." It is "far superior to most science fiction in creating a world that could be true. His is the sort of story I could believe because it is subtle, detailed, and restrained, involves only a very limited government conspiracy, and does not digress into any kind of speculation."

It was just these qualities about the tale, PsychoSpy noted, that explain why it "appeals to engineers, computer programmers, and other techie types." It is "heavy on plausible technical details and free of the emotional overtones" that characterize many shrill UFO accounts. "If Lazar's story is fiction, it's great fiction, filled with a richness of plausible details and complex philosophical dilemmas that you can't find in most popular novels these days."

It was exactly this similarity to a fictional character's tale—sometimes detailed, sometimes vague, highly subjective, with even a hallucinatory quality, the sense of an imperfect memory washed out by mind control or other means—that made me, too, think of Lazar as a fictional character.

For me, the weirdest part of the story was not the saucers or the aliens. It was the poster that Lazar said he saw in the offices at S-4, the one with the picture of a saucer hovering above the desert and the words, "They're here!" It looked, he said, as if it had come from Kmart.

The more I studied his tale, the more Lazar reminded me of the antihero of a science-fiction novel by Philip K. Dick. Many of Dick's protagonists are dweebish, sometimes seedy, average guys who get caught up in matters of planetary import. They live in crass commercial worlds while dealing with what they consider important philosophical questions. And they face realities that fade in and out of each other, raising larger questions: Are there sinister influences at work or only demented solipsism? Is it in my head or is something very wrong with this universe? They often feel they are in a carefully crafted illusion, but that some of the workers have spoiled the effect by leaving empty sandwich wrappers and soda bottles around.

Lazar's tale has this same quality of a half-waking dream. Levels of reality drift in and out of each other in a strange but compelling way. Details of the quotidian world blend with those of the Lore.

Lazar, for instance, noticed that the security badges bore blue and white stripes and the legend "Majestic." "It made me crack a smile," he commented, because "Majestic" is straight out of the Lore: MJ-12, which stood for either Majestic or Majic 12, is the famed and much-debated secret committee in UFO legend charged with recovering and hiding flying saucers. MJ was said to be a security clearance "38 levels above Q," or top secret.

"I don't know whether it was a kind of nostalgia thing," he commented. "I began to wonder is this really the Majestic everyone talks about, or was it something done almost for nostalgia reasons? . . . Assuming the Majestic 12 documents were false, did these guys just use this insignia for the hell of it, kind of as a joke?"

The flip side of Lazar's unwillingness to speculate is that the big issues raised by saucer lore are ignored: How did we get the saucers? Do the aliens run the base? Was there really a link to Roswell or to MJ-12? Unanswered questions lie heavily over the Lazar story and provide much of its fascinating quality, but the gaps in the tale could also be designed to make it easier for true believers to link it to their own wider conclusions.

Another dreamy effect is the strange alien book Lazar says he saw at S-4. Its pages were translucent, like a series of acetate layers, so that you could see into a house, X-ray style, from shingles to framing to chimney inside. He was allowed to read the book, which combined a history of the earth and a history of a planet in the star system Reticulum 4, where the saucers originated. Human beings are referred to as "containers"— for souls or for genes or whatever is unclear. (The term "containers" caught the imagination of UFO buffs; the Heaven's Gate cult would use it in their teachings.) Some sixty-five "genetic interventions" beginning in the epoch when men were still apes were described. "Intervention" seemed to Lazar to mean manipulation of DNA, and appeared designed to make humans a breeding species for the aliens—a kind of grafting stock for a race that had lost its ability to reproduce.

The book serves as a means to introduce much more information than would have come to Lazar's attention directly, but it seems a clumsy plot device, worthy of a computer adventure game or a wavy transition in a film from an opening book to real action.

Unlike many UFO sources, Lazar had begun as a skeptic; he had gone on record as deriding the youfers. A convincing detail is Lazar's statement that when he first caught sight of the saucers, he thought they were terrestrial military craft. "Well, there's the explanation for UFOs," he thought. We must have made them. But when he learned they were not from Earth, he had a strange reaction. That night, he said, he lay in bed, giggling, unable to sleep. Lazar had a charming reluctance to overstate. "I hate to mention this," he'd begin. "I don't want to get too deeply into that," he would say in answer to a question, or "I don't like to talk about this." He was almost coyly casual about his one sighting of an actual alien. It could have been a mannequin, he says, or a mock-up. "It could have been a million things."

But Lazar's story has the useful feature, too, of suggesting associations with the rest of the Lore, like the round tabs of jigsaw-puzzle pieces. The very vagueness and limits of his knowledge inspires listeners to make their own links. He's not sure where the saucers he saw came from—could it have been Roswell, or the storage site at Hangar 18 at Wright-Pat? He hears rumors of a shoot-out with aliens—perhaps it was at S-4, or Area 51, or at Dulce, as the Lore tends to have it?

A recurrent theme in Lazar's story was his feeling that his employers at S-4 were "trying to make him disappear" by removing records. This happened even before he left the job. He claimed it was this sense that he was being made invisible that led him to go public. He couldn't find records of his own life, he said. "They're trying to make me look nonexistent," or, in an oddly dislocated locution, he felt "that someone was going to disappear."

Worse, he was forgetting things. Had they done something to his mind? Had he been given something to drink, as the Lore held they often did to interlopers (it was supposed to smell like Pine-Sol)?

His memories were disappearing, too. By September 1990 he was complaining he had forgotten the name of the two modes of travel of the saucers—one low speed, the other high, intergalactic speed—and resorted to calling them alpha and beta. Nor could he any longer remember an important coefficient for one of the processes or certain frequencies of the gravity wave and other details he was convinced he had once known. "I've developed a mental block," he said. "It really bugs me." He went to a hypnotherapist to help him remember, but it was not very successful.

Lear noticed that Lazar had begun to forget things. "Don't you remember that night you came over to my house all excited?" Lear asked him. Lazar had completely forgotten.

One night in December 1988, or January 1989, Lear recalls, Lazar came by his house in a state of high excitement. It was bitter cold, "but we talked outside because it made him more comfortable. He was in shirtsleeves. He told me about seeing the alien. He was very excited. Now, he can't remember it."

"I saw a disc," Lear says Lazar told him.

"Ours or theirs?"

"Theirs. I just got back from the test site."

"Oh my God. What are you doing here? You should continue to work up there for a while. Don't jeopardize your security clearance."

"But, John," Lazar replied, "you've taken so much flak about this stuff that I'm going to tell you."

"And for the next three hours and forty-seven minutes he proceeded to tell me all of it. He told me we did have secret bases on the moon and Mars. He told me things, some of which were so unbelievable [that] had I not known Bob I would have been very suspicious."

————

Once Lazar was asked, "Don't you feel—no pun intended—alienated? In fact aren't you kind of connected with them, and removed from the rest of society that doesn't accept that?"

"Absolutely," he answered. "I feel like I really know what's going on, and everyone's an idiot. I really feel that way. Alienated is the perfect word for it." He was, you might say, a classically alienated type. But the S-4 experience had given order to his life.

The saucers, he said, "made it all make sense. It's the only thing that makes sense. It takes a lot of the confusion out of things. A lot more knits together . . ."

Still, as PsychoSpy had urged in his essay "Lazar as Fictional Character," consider Lazar's story as story. He implies that if Lazar did not exist, the youfers would find it necessary to invent him. That they may have invented him, or that the Air Force Office of Special Investigations (AFOSI) or some other government organization may have invented him, or that he invented himself—all are possibilities that hang in the air like the lights over Dreamland. But who would invent Lazar, and why? Was he a government disinformation agent? Why? As cover for secret programs? To make sure that people believe the lights they see moving above the Jumbled Hills are flying saucers instead of manned terrestrial aircraft or, more likely, unmanned aerial vehicles (UAVs)? The Stealth fighter was revealed in the autumn of 1988 just as Lazar went to work at S-4. Was there a connection? (Indeed, the first images of the Stealth fighter, heavily airbrushed, were released about the time Lazar surfaced.) But Lazar's story would only draw more curious viewers to the perimeter, where they might see real aircraft while looking for Lazar's saucers.

Did Lazar create Lazar? For money or for fame? There was a film deal with New Line Pictures, although the amount of Lazar's income from the rights was unclear. The film languished in production. Originally due in 1994, it went through many scripts and suffered from troubles at New Line. He had been paid to serve as consultant for a plastic kit of the saucer "Sport Model" for the Testor company. Packed with each kit was a poster, just as Lazar had described, bearing the words "They're here!" Was his goal to become a legend in his own mind, to feel comfortable and real there in front of the blackboard in the pose of Teller or Oppenheimer, to become at last a real authority?

Lazar's story hovers about the Ridge. Chewed over, tugged at, poked, prodded, and twisted, it quickly became a modern legend, obsessing viewers who came to the Black Mailbox to see if they could see Lazar's saucers.

4. Aurora

It was cold, Chuck Clark told me, sitting across the table at the Little A"Le"Inn, twenty below, when he saw the Aurora. "But I served in Korea where it was colder than that all the time. I'd been waiting for hours and only saw it for a few seconds, silhouetted against the light when they pulled the hangar door open. It rolled out and the door closed and it took off." What shape was it? He was vague. "But many times I've seen the blue flame of the methane engines on the test stand behind the hangar"—the big hangar, the one the youfers called Hangar 18. "It exists."

Aurora, the most mythical of the planes above Dreamland, was believed to be the successor to the Blackbirds, a mother-daughter ship arrangement, flying at Mach 8, perhaps the craft that leaves the little putt-putt doughnut-on-a-rope contrail.

———

In the 1890s, strange reports began to surface of mysterious airships drifting over the Midwest and West. They were heard in Appleton, Wisconsin, and Harrisburg, Arkansas, but Texas had the largest number of reports. Some of the crews talked to people on the ground. One group asked for food. Another was said to have sung "Nearer My God to Thee." There was even a cattle mutilation report: A steer had been lassoed, pulled into the airship, roasted, and eaten, with only the skin and bones dropped back overboard.

The reports resembled those of the post–World War II flying-saucer era, except that the speeds cited were tens or hundreds of miles per hour, rather than hundreds or thousands, and the materials described were no more exotic than aluminum.

The mid-1890s were a period of economic depression, political instability, and general cultural unease. The first dirigibles—"airships"—had flown in Europe, and Samuel Langley of the Smithsonian Institution flew his crude aircraft from a houseboat on the Potomac River in May 1896 and garnered widespread publicity. The invention of the airplane seemed imminent.

The height of the craze came in April 1897. One report, in the *Dallas Morning Times*, on April 19, came from the small town of Aurora, Texas. Aurora was then a dusty little town, and having an airship sighting meant being up to date; an account of another sighting in Denton, Texas, had suggested to the local newspaper editor proof that Denton "was not behind" other towns. The *Morning Times* told of a craft crashing into a windmill, of wreckage and a pilot's log. There was speculation that it was from Mars and even word that one of the crewmen was killed in the crash and buried in Aurora.

The Dallas newspaper's report made startling claims:

About 6 o'clock this morning the early risers of Aurora were astonished at the sudden appearance of the airship which has been sailing throughout the country. It was travelling due north, and much nearer the earth than before. Evidently some of the machinery was out of order, for it was making a speed of only ten or twelve miles an hour, and gradually settling toward the earth. It sailed over the public square and when it reached the north part of town [it] collided with the tower of judge Proctor's windmill and went to pieces with a terrific explosion, scattering debris over several acres of ground, wrecking the windmill and water tank and destroying the judge's flower garden. The pilot of the ship is supposed to have been the only one aboard, and while his remains are badly disfigured, enough of the original has been picked up to show that he was not an inhabitant of this world.

Mr. T. J. Weems, the U.S. [Army] Signal Service officer at this place and an authority on astronomy, gives it as his opinion that he [the pilot] was a native of the planet Mars. Papers found on his person—evidently the records of his travels—are written in some unknown hieroglyphics, and cannot be deciphered. This ship was too badly wrecked to form any conclusion as to its construction or motive power. It was built of an unknown metal, resembling somewhat a mixture of aluminum and silver, and it must have weighed several tons. The town today is full of people who are viewing the wreckage and gathering specimens of strange metal from the debris. The pilot's funeral will take place at noon tomorrow.

Signed: E. E. Haydon.

No one at the *Morning News* picked up on the dispatch's dramatic suggestions. None of the strange metal ever showed up; the "papers" were not shown. And no one inquired about the pilot's grave. But the account was a foreshadowing of a Roswell-style crash—the hieroglyphics, the widely scattered debris, the strange materials, the recovered body, were all standard elements of twentieth-century saucer crashes.

The story was not taken up again until 1967, in an account in a British UFO publication by Jacques Vallee and Donald B. Hanlon called "Airships over Texas." After that story appeared, a UFO investigator visited Aurora. He found that the Proctor farm where the crash had been reported was now a gas station run by a man named Brawley Oates. Oates referred the investigator to another man, Oscar Lowry, who had been eleven at the time of the incident.

Lowry and other surviving witnesses strongly suggested that the whole thing had been a hoax. There was no Army Signal officer in the town—T. J. Weems was the town blacksmith. Proctor's farm didn't even have a windmill.

E. E. Haydon, the stringer for the Dallas paper who wrote the story, was the local cotton buyer. He had noted the decline of Aurora since a new railroad had bypassed the town. The story was almost certainly a prank, in the spirit of Rachel's efforts to cash in on local UFOs.

In 1973, with the country sitting through the Watergate hearings and, perhaps not incidentally, finding itself in the grip of one of its periodic waves of UFO sightings, reporters from U.P.I. picked up on the old Aurora tale. A report that appeared in many newspapers on May 24, 1973, quoted Hayden Hewes, director of an organization called the International UFO Bureau, who had gone to Aurora to investigate. Hewes claimed to have discovered the spaceman's grave and threatened to go to court to have it opened. He found a strange rock marked with an arrow and three circles in the cemetery and reported that the spaceman had been buried under it.

Reuters and the Associated Press joined the chase. The A.P. reported that samples of strange metal had been found near the gas station. When analyzed, they turned out to be mundanely terrestrial pot metal. Reuters interviewed a ninety-one-year-old woman who claimed to recall that the pilot had been buried in the cemetery, which was run by the local Masonic order and an organization called the Aurora Cemetery Association. But the association's map of the cemetery plots revealed no sign of the spaceman's grave or of any unidentified graves. The group blocked attempts to dig up the place, and on the night of June 14, 1973, the strange rock disappeared as mysteriously as it had arrived.

———

The name Aurora returned in the 1980s linked to the most mysterious of mystery airplanes. What the Lazar story was for UFO watchers, Aurora was for black-plane buffs. In the late eighties and early nineties, Aurora became the focus of speculation among the watchers—the pinup goddess of the Interceptors. The name evoked high-flying associations: Aurora, goddess of the dawn, or aurora borealis, the northern lights that sometimes so enraptured pilots they would fly toward them to their deaths.

The word *Aurora* entered the lore of black aircraft when it popped up in a P-1, or procurement budget document, near line items for the U-2 and the SR-71, and attached to the phrase "air-breathing reconnaissance." Its inclusion appeared to be a mistake, but the stealthies and Skunkers noticed it. And they noticed the next year when the size of the requested appropriation for Aurora for fiscal 1987 rose from $8 million in fiscal 1986 to $2.3 billion. The next year the item vanished. They assumed it was a successor to the Blackbird and the legendary U-2. The Skunk Works must be at it again.

The first reports came in the aviation press. And in 1988 *The New York Times* ran a story on the plane that claimed it could fly as fast as Mach 6.

In 1989 an oil-drilling engineer named Chris Gibson spotted what may have been Aurora refueling with two F-111s. Gibson, perhaps a bit too conveniently, skeptics noted, was a member of the Royal Observer Corps, trained in recognizing aircraft. In August 1989, Gibson told me, he was working on a petroleum drilling rig called the Galveston Key in the Indefatigable oil field in the North Sea. He was below decks when his coworker Graeme Winton came down and told him to hurry above.

"Have a look at this," Winton said, pointing out a group of planes flying overhead: a large one, two smaller ones, and a strange triangular one.

After a while, Gibson explained to Winton, aircraft observers count on an almost subliminal feel for the shape or gestalt of an aircraft, called the "sit," similar to what bird-watchers refer to as "jizz." But no "sit" seemed right for the triangle.

"The big one is a KC-135 Stratotanker," he told Winton. "The two on the left are F-111s, and I don't know what the fourth is."

"I thought you were an expert," Winton commented.

"I am."

"Some expert."

At first Gibson thought the triangle might be another F-111, but there were no gaps in its wings and it was too long. The F-117 had just been made public, but the triangle was too big for one of those. Nor was it a French Mirage IV fighter. Gibson was stumped.

Back in his quarters, Gibson consulted the aircraft recognition manual that he considered the best in the world: the Flykendingsbog, published by the Danish civilian spotter group, the Luftmelderkorpset. But no plane in the book looked anything like what he had seen.

Gibson then made a drawing of the triangular craft and sent it to several aviation journalists, including the highly respected Bill Sweetman, who much later presented the sighting, in Jane's, the aviation publication, in December 1992, as one of the linchpins of a pro-Aurora argument. The plane, Sweetman concluded, could fly at Mach 8, reaching anywhere on Earth within three hours. It had first taken to the skies, he believed, in 1985, at Groom Lake, and likely flew in and out of Machrihanish, the Scottish special forces base that had also hosted the SR-71.

In 1990, after a ceremonial flyover above the Lockheed Skunk Works, which the ailing genius Kelly Johnson viewed from his car, the SR-71 was retired, lending strength to the Aurora stories. The Air Force or CIA wouldn't have retired the Blackbird, the reasoning went, if they didn't have something else ready to replace it. Why had the Air Force not fought harder to keep the SR-71?

Complex politics swirled about the SR-71. While the Blackbird lacked powerful patrons within the Pentagon, its legend attracted many in Congress, which several times had restored the Blackbird to the budget after the Air Force had removed it. Aurora seemed the logical next project for the Skunk Works, a plane that flew higher and faster than any then known, kept under wraps as long as possible.

Aurora, the story soon came to include, was powered by methane, a technology involving cryogenics, which the Skunk Works had explored as early as 1957. At that time, it had nearly built the hydrogen-powered CL-400 or Suntan, but Skunk Works boss Kelly Johnson killed the project at the last minute when he realized the prohibitive cost of setting up an infrastructure for handling liquid hydrogen at bases around the world and refueling in flight.

Liquid methane might work better. It might power an Aurora that girdled the globe, a recon plane, but one that might also be able to drop a wicked heavy projectile on a hardened command post with an uppity dictator inside it. Johnson had advocated such a system years ago, using the SR-71. Dropped while flying at such speeds, a heavy hardened-steel projectile is like an A-bomb—each thousand miles of velocity is worth a pound of TNT.

In 1991 a series of "skyquakes," as the local media liked to call them, long rumbling sounds, rolled over Los Angeles. To seismologist Jim Mori, these suggested the sonic booms of a craft returning from an altitude of, say, 100,000 feet, or even from space, descending over L.A. to land in Dreamland.

Sightings around the same time in Palmdale and the Antelope Valley proliferated. Many of the reports depicted a long triangular craft, with wings swept back about 70 degrees. Others suggested an XB-70-like craft, or a "mother ship" carrying a smaller "daughter" craft on its back.

A TV writer named Glenn Emery reported a sighting in May 1992 near Atlanta, hardly black-plane country. In August 1992 more reports surfaced of delta shapes. The sound described in several reports, including one near Mojave, California, was a "low-pitched rumble." That month, a viewer near Helendale, California, location of Lockheed's radar cross-section (RCS) test facility, described a craft crossing the road at an altitude of less than two hundred feet. It may have landed at Helendale, the reports said, because the Groom and Nellis areas were covered with severe thunderstorms.

There were reports of shrouded shapes being loaded onto cargo planes at the Skunk Works in Burbank and of airliners in near misses with strange craft. Airline pilots reported several near misses with triangular craft.[1]

In August 1992, John Pike and the Federation of American Scientists (FAS) published their *Mystery Aircraft* report, which took at best an agnostic view. The study pointed out the epistemological problems: There were too many sightings, too much information, too many possible planes—and yet not enough evidence. And despite the budget document listings, the FAS report pointed out, no money had ever actually been appropriated for the Aurora item before it was removed.

As usual, the signal-to-noise ratio was invoked. Based on the report, *The New York Times* came out with a story in January 1993 that denied Aurora's existence. But Aurora flew on. At least on aviation and popular science magazine covers, it flew with all the fidelity skilled airbrush and gouache could convey. The paintings and models made the near mythical craft seem as real as any Piper Cub at the local landing strip—or, rather, more real. Amphibian, feline, raylike shapes, delicately modeled, seen against orange sunsets and blue depths of sky—if they did not exist they should have.

In 1993 the Testor company released a model John Andrews had designed. It adopted the theory that Aurora was a "mother ship" with a smaller vehicle on its back. The mother ship bore the name "SR-75 Penetrator," and on its back rode the "XR-7 Thunderdart." The Thunderdart was supposed to fly at Mach 7 and boasted the pulse detonation wave engines that emitted the already famed doughnut-on-a-rope contrail.

The model made the idea of Aurora inescapable. Such craft should exist whether it did or not. It was hard not to believe in a craft that someone had so carefully and thoroughly imagined, designed parts of, and written instructions for that that read like this: "Podded Engines. Assembly. 1. Cement centerbody vane, 71 G, to centerbody wall and vane, 72 G. Now cement the vane/wall unit into the center spike, 73 G. Now cement the centerbody flow ring, 74 G, to the centerbody."

Jim Goodall, aviation journalist and black-plane expert, was convinced. Goodall believed that about $15 billion had been spent on the thing, that it was there to sniff out Third World nukes, a joint project of the United States and the former Soviet Union.

Even Bob Lazar claimed to have seen what he thought was Aurora, inside Dreamland.

———

Speculation over Aurora brought all sorts of proposed hypersonic craft designs out of the closet as stealthies rushed to find corroboration for a real plane. These were dream wings, paper airplanes. Aircraft companies and engineers are constantly dreaming up possible airplanes. Sometimes they are simply fantasies, aeronautical engineers' wet dreams, and sometimes they are teasers, like concept vehicles shown at car shows, intended to whet the public's appetite and that of the generals in the Pentagon.

By the fall of 1993, Bill Sweetman had written a book on Aurora, consisting mostly of citations of these earlier hypersonic aircraft proposals, going back to the early supersonic X planes. There were dozens of them, pictured with slick contractor illustrations of lifting bodies and wave-riders (triangular aircraft that surf on the shock wave produced when they push beyond the speed of sound), many of them intended to be launched from the back of another aircraft. Also included was Lockheed's hypersonic glide vehicle, which was designed to reach Mach 18.

One version of the Aurora story held that work began in 1983 to create a successor to the SR-71. It was called Q, *Aerotech News* reported, from "quantum leap" in technology, but it had become too expensive and was canceled. To Jim Goodall, cost was no problem. The airplane would cost, say, a billion dollars a year. What airplane didn't cost that much? he argued. The number was easy. And it was easy to hide that much.

Black-budget watcher Paul McGinnis, known as Trader, at first believed that Aurora was a program code-named Senior Citizen. But he tracked that one down and concluded it was a stealthy transport—a short-takeoff-and-landing craft for sneaking troops behind enemy lines. Later, the program he finally decided was the real Aurora was one he knew only by the budget-line code number 0603223F.

In another theory, Aurora was not hidden at all, but was the shadow of Ronald Reagan's "Orient Express," a supersonic dream plane that would fly from New York to Tokyo in a couple of hours. No one could figure out how this projected passenger craft, formally called the National Aerospace Plane (NASP), made any economic sense. John Pike speculated that Aurora might be hiding in plain sight as the NASP—"a purloined letter" of an airplane, Pike called it. Sweetman noticed, too, that the NASP planners were confidently counting on building the Orient Express from a titanium alloy that had never been used before—at least in any publicly known aircraft.

Was it conceivable that Aurora was not a manned airplane, but a robotic one, an unmanned aerial vehicle, rumored to be called Q or Tier III? Perhaps the romance of the name was elusive as well: The "Glossary of Aerospace Terms and Abbreviations," a supplement to the aerospace magazine *Air International*, claimed that Aurora was an acronym for "AUtomatic Retrieval Of Remotely-piloted Aircraft."

One theory held that the plane had been canceled in 1986 because it was too expensive or didn't work. Another said it had suffered catastrophic failure on the eve of the Gulf War. Yet another reported that it had been pushed ahead because of the Air Force's desperation for a space plane in the wake of the *Challenger* disaster in January 1986 and the failure of two Titan booster rockets carrying spy satellites. But the dates did not jibe with the budget document. The B-2 had been given the go-ahead in 1981, and did not fly until 1989. How long would it take even the Skunk Works to bring an Aurora to fruition?

Supporting the "it was a bust" theory was a report in July 1994 by the Senate Appropriations Committee stating that "The system which some hoped would be developed and procured as a follow-on to the SR-71 has not materialized."

The myth of the airplane came to resemble its possible namesake, the aurora borealis. It suggested a shimmery, elusive veil of rumint, charging the imaginations and dreams of the Interceptors and stealth watchers.

———

After Sweetman's report in *Jane's*, which *The Wall Street Journal* and *The Washington Post* picked up, the government responded. Donald Rice, Secretary of the Air Force, issued a categorical denial in a letter to the *Post* in December 1992.

"Let me reiterate what I have said publicly for months," he wrote.

The Air Force has no such program either known as "Aurora" or by any other name. And if such a program existed elsewhere, I'd know about it—and I don't. Furthermore, the Air Force has neither created nor released cover stories to protect any program like "Aurora." I can't be more unambiguous than that. When the latest spate of "Aurora" stories appeared, I once again had my staff look into each alleged "sighting" to see what could be fueling the fire. Some reported "sightings" will probably never be explained simply because there isn't enough information to investigate. Other accounts, such as of sonic booms over California, the near collision with a commercial airliner, and strange shapes loaded into Air Force aircraft, are easily explained and we have done so numerous times on the record. I have never hedged a denial over any issue related to the so-called "Aurora." The Air Force has no aircraft or aircraft program remotely similar to the capabilities being attributed to the "Aurora." While I know this letter will not stop the speculation, I feel that I must set the record straight.

The Air Force commissioned an independent testing lab to show that the "skyquakes" in Los Angeles were nothing more than booms from offshore Navy fighters. Yet whether it was due to Rice's denial or the arrival of a new administration, skepticism in the press began to grow.

In Amarillo, the arch-Interceptor Steve Douglass had scanned a revealing conversation from an Air Force aircraft phone. The transmission took place on the "Mystic Star" network used by aircraft transporting heads of state and military VIPs, including Air Force One. The transmission was made in the clear on December 10, 1992, when a general placed a phone-patch from SAM (Special Air Mission) 204 through Andrews AFB to "AF public relations."

Aurora was discussed. The general quoted the article in The Washington Post as well as the one in Jane's. He said, "It's almost laughable the number of hokey inputs they had. It's kind of similar to the UFO flap. We need to develop a release in response to inquiries. The guts of this should be that we've looked at the technical aspects of the sightings and what the logical answers for them are. You can quote Dr. Mori and cite the Lincoln Lab physics and the FAA's efforts to debunk other incidents. Go through three or four of the sightings, take each one on and conclude with a paragraph that says the fantasy of Aurora doesn't exist."

They went on to discuss the sighting in the North Sea from an oil drilling platform. "Someone saw something accompanied by three F-111s. The secretary wants us to say it was an F-117."

To Steve, there was clearly a cover-up under way.

———

Aurora vanished from the next round of budget documents, and Ben Rich would later report that Aurora was the code name for the funding of the B-2 competition between Lockheed and Northrop. Others in the industry made fun of the legend. A stealth expert at Northrop once asked me, "Have you heard the news about Aurora?" He waited the requisite two beats and then said, "It's an Oldsmobile."

And true enough, Oldsmobile had come out with a dramatic-looking new car named Aurora (the designer credited the F-15 as one inspiration for its shape) that was supposed to help the company's laggard sales. The ads for the Oldsmobile even referred to the airplane: You can't see the Air Force's, they said, but you can buy ours.

———

In 1985 a movie loosely based on the 1897 Aurora, Texas, airship story appeared. It featured an elfin ET who wore jeweled, medieval clothing and piloted a Victorian flying saucer amid sets left over from a cheap Western. The spaceship, its rivets exposed like Captain Nemo's *Nautilus*, suggested an 1890s and not a 1980s version of high-tech.

"It was a squatty shape with wings," says the movie's colorful old coot, who makes patent elixir, "but the strangest thing was the little feller driving it."

The film spins out the original story: The landing is real, and a newspaper editor capitalizes on it to save her ailing publication and bring the town fame.

At the time of the great airship wave, William Randolph Hearst denounced the reports of the sightings in the same tones that future newspaper editors would use for castigating tabloid newspapers. In a *San Francisco Examiner* editorial of December 5, 1896, Hearst intoned: "Fake journalism has a good deal to answer for, but we do not recall a more discernible exploit in that line than the persistent attempt to make the public believe that the air in this vicinity is populated with airships. It has been manifest for weeks that the whole airship story is pure myth." It was a shrill tone to take for a man who, two years later, would be largely credited with puffing up tensions in Cuba that propelled the United States into the Spanish-American War.

——

Was the latter-day Aurora a headline without a war? Or could Aurora have been as mythical as the long-ago airships over Aurora, Texas? A craft full of hot air, a shape compounded of disinformation?

Significantly, Aurora as an imaginary aircraft could have had some of the effects of an actual plane. It could, for example, have made potential enemies aware that they could be observed at any moment. Did whoever named the craft Aurora know about the Texas town and its tale? Was this an inside joke, deliberate political disinformation?

By the mid-nineties a flock of new high-speed aircraft came into the open. One was called LoFlyte, a so-called waverider. And when Lockheed Martin received a contract to build the X-33, the hypersonic suborbital aircraft, Skunkers became suspicious. The promised delivery date and comparatively low bid suggested that Lockheed had technology already available—possibly from Aurora—to give it a head start. Was the X-33 simply the "white" version of Aurora?

If there was no Aurora, or nothing like it, why were buildings going up so fast at Area 51? Why were Wall Street analysts pointing to large, mysterious sources of income in Lockheed's annual reports? Why were Lockheed's parking lots full? What was it that needed a six-mile runway across Groom Lake, in Dreamland? Questions like these, as much as the tales of Lazar's saucers, drew the curious in greater and greater numbers to the perimeter of Area 51.

5. Maps

The Little A"Le"Inn did a good business in maps—bought from the government and significantly marked up. Naturally they did not show the base over the Ridge.

The fascination Dreamland radiated began with the fact that for years it did not officially exist. A map I bought at the Bureau of Land Management office in Las Vegas did not show it. The 1:100,000 metric scale, 30 x 60 minute map from 1985 claimed to display "highways, roads and other man-made structures" but bore no signs of runways, hangars, or the buildings that housed hundreds of workers and engineers at the base. But why should Dreamland be on the map? It was after all not a real but an imagined place, a virtual landscape, a "notional" land, and its map was to be found drawn not on the ground but on the mind.

Groom Lake and Dreamland were part of a wider map of secret facilities, mystery spots that represent a significant portion of tax dollars at work: air bases and test sites, controlled airspaces and anonymous buildings housing research facilities. It belongs to the same cultural landscape as the nuclear labs in Los Alamos and Sandia, New Mexico, the Blue Cube in Sunnyvale, California, which controls spy satellites, the CIA training facility at Camp Peary, Virginia, and the National Reconnaissance Office's headquarters outside of Washington, D.C.—the "stealth building" kept secret from Congress even while under construction. Many of these facilities make up the Southwest Test and Training Range Complex, which runs from White Sands and Fort Huachuca in the south to the Utah Test and Training Range in the north. Included are the Air Force Flight Test Center at Edwards, the best known, most open of the areas, even including the closed-off "North Base," and China Lake, the Navy's radar and electronic test site to the north and east of Edwards. There is a difference, though: All these other facilities have long been acknowledged.

Dreamland had no edges—the Minister's phrase kept coming back to me. But it had ties and umbilicals to Edwards, to the contractors in Las Vegas, to the Air Force labs at Wright Patterson in Dayton. The ties reached all the way back to the Pentagon, whose shape has transformed from that of an old star-shaped fort into an icon of the new military-industrial complex. I thought of the whole network of facilities that were kin to Dreamland as a mysterious distant land: Pentagonia, marked with its own patterns, somehow similar to the Dreamings of the aboriginal peoples of Australia; or as an expanding metropolis, a ghost metro area with its own suburbs, industrial parks, malls.

Dreamland was born of the culture of secrecy; its owners—the Department of Energy and before that the Atomic Energy Commission, on the one hand, and the Air Force on the other—sat at the junction of the twin ideologies of nuclear power and airpower. This invisible culture cast a great shadow, which was the culture of ufology—the antimatter of the matter.

I tried to create a mind-map of Dreamland, based on certain marketing presentations I had seen. A car maker, for instance, might mind-map the image of a vehicle. One axis—latitude—would mark "sporty" versus "practical" while another—longitude—might distinguish a range of impressions from "luxury" to "basic transportation." It seemed to me you could mind-map the cultures of the world of nukes and the world of airpower in a way that could neatly correspond with the overlaps of the test site and the Nellis range on the physical map.

In time I tried to map, too, the mind-sets of those on the Ridge: "Believe secret airplanes are being tested" on a line with "Believe alien technology is being tested." The youfers and the Interceptors would be at either pole. In the middle were a surprising number of people who bought into both—and even more who were simply tempted to believe.

The other axis would distinguish "those who think it ought to remain secret" from "those who want it opened as much as possible." Oddly enough, by the mid-nineties I was hearing the youfers cluster at the first end of the axis, with Huff and Lear saying they now thought it should all remain closed, the whole story kept down because people weren't ready, couldn't deal with it.

The closer you got, the harder it was to see. It became a cliché that everyone saw from the Ridge what they wanted to see: "I wouldn't have seen it if I hadn't believed it." To see the whole thing you had to step away, and look from many perspectives, through many eyes.

Still, I studied the mint and mocha shades of the Coast and Geodetic Survey maps and looked at the official tourist map of Nevada, with its upbeat registration of ghost towns. I put my hands on maps from the Defense Mapping Agency as well as the Federal Aviation Administration aerial charts with their landscape of ochers and burnt yellow hatched with the purple edges of restricted military operations areas like blackberry juice stains or old, fading bruises. The signal stations—VORs—for aviation guidance were rendered as gear-toothed compass wheels.

—————

One afternoon I stopped by the state historical museum in Las Vegas. There was a display on the Shoshone tribes who originally lived in the area. Beside a panoramic photograph of Tonopah, the town north of Dreamland proper, in the heyday of the silver boom—a collection of mine tailings and shacks and a hotel bearing a Bull Durham ad—hung a map promoting the Tonopah and Tidewater railroad, the brainchild of Francis "Borax" Smith. Smith replaced the famed twenty-mule teams hauling borax from the mines around Trona with trains, and they still ran. In Mojave I had many times heard the Trona train rumbling through in the middle of the night, a seemingly endless succession of low dark ore cars coming in from the northeast.

The landscape of nearly a hundred years ago looked more inhabited and detailed, packed with mines and claims and crisscrossing railroads. In the map's legend, the twin T's of the railroad name were cleverly eye-punned into twin T-rails. I was shocked to see no boundaries across the map—no dotted perimeters, no shaded restricted areas, no overlapping colors—so used had I become to maps of restricted spaces.

I looked at every map I could find. I even "flew" over the lake and the mountains on a CD-ROM map that could show any part of the landscape of the country in three dimensions. I flew over the mountains from the area of the Black Mailbox, moved up the Groom Road, then over the hills, zooming along the runway and past the hangars, neither of which were marked, and turning to cross over Bald Mountain with a sickening plunge like a roller coaster's. I turned on the terrain-following feature and, nosing down, saw it all dissolve as proximity overwhelmed the program's resolution and individual pixels grew into colored angular shapes, into facets like those of a stealth plane. Finally, the screen turned as blank as the maps were before the miners and military arrived.

———

What the maps did show was that Dreamland is a place where things overlap. Mojave Desert meets Great Basin and quartzite overshoots Cambrian limestone; the range of the ancient Anasazi fades into and over that of the Fremont culture, where Nevada Test Site overlaps the Nellis Air Force gunnery and bombing range—which in turn are overlapped by the National Desert Wildlife Range, created in 1936 by FDR to save the bighorn sheep.

The signs warning of use of deadly force on Dreamland's perimeter refer to the USAF/DOE liaison office in Las Vegas, for which they provide a post box number. By the best accounts, the Air Force and Department of Energy jointly administer the area, under a "Memo of Understanding."

The Atomic Energy Commission took control of the area just to the south and west of the dry lake in 1950. Airspace here was limited beginning in 1955, and the area formally shifted from the public lands of the Nellis range to the control of the Atomic Energy Commission.

Nellis Air Force Base had greater needs, too, and by 1959 all the grazing and most of the mineral rights within the range were purchased by the Air Force.

In 1956, 369,280 acres of the Nellis range to the northwest of the lake were lent to the AEC as the Tonopah Test Range for ballistic missiles. In 1958 Public Land Order 1662, signed by one Roger Ernst, assistant secretary of the interior, withdrew from the public lands 38,400 acres (60 square miles) for use "by the Atomic Energy Commission in connection with the Nevada Test Site." The area was the first formal survey of the six-by-ten-mile "box" around the base.

On August 11, 1961, with tensions rising in Berlin and bad news from Laos, the FAA established a new restricted airspace, designated R-4808 and covering the test site and Groom Lake. On thousands of bulletin boards in large airports and tiny control towers across the country, a NOTAM—"Notice to Airmen"—apprised pilots of the new boundary. In January 1962, the restricted airspace was expanded to 22 by 20 nautical miles in response to a request by the Air Force citing "an immediate and urgent need due to a classified project." By the early sixties, military maps began to show the air controllers' name for new restricted airspace over and around the base. Bordering airspaces known as Coyote, Caliente, and Alamo was "Dreamland."

————

Starting about 1978, "in the interest of public safety and national defense," the Air Force began—and here the authors of the 1985 Environmental Impact Statement for the Area 51 region become gloriously politic and delicate—"actively discouraging, and at times preventing, public or private entry to the Groom Mountain Range." The government also put up fences on the east side of the range.

The next seizure, under Public Law 98-485, in October 1984, included Bald Mountain, the nine-thousand-foot former volcano. In a letter dated July 6, 1984, Deputy Assistant Secretary of the Air Force James Boatright assured rancher Steve Medlin, the owner of the Black Mailbox, of his continuing water and grazing rights. These are measured out by the BLM in Animal Unit Months (AUMs). The Bald Mountain Allotment contains some 5,811 AUMs, which translates to 480 head of cattle and five horses. They assured the Sheahans, the owners of the site, of continued access to Groom Mine. But the Sheahans, heading there one day, found the way blocked by blue-bereted Air Force police.

Dreamland and the adjoining nuclear test site had become a de facto nature preserve. Animals could move back and forth between the two in a way humans could not. In the spring of 1985, when environmentalists visited the area to support the Air Force's effort to withdraw from public use additional land around Groom Lake, they found that wildlife was flourishing. Jackrabbit and cottontail were abundant, as were coyotes, mule deer, badger, and kit foxes. Two mountain lions were recorded. The chukar partridge had been growing in numbers.

The area is home to six kinds of rattlesnakes, the ferruginous hawk, Swainson's hawk, mountain plover, western snowy plover, and long-billed curlew, as well as four species of bats, ranging from the little brown myotis to Townsend's big-eared. Naturalists defined several plant and animal communities in the area, ranging from saltbush to mixed Mojave, blackbrush/sagebrush to pinyon/juniper to mountain mahogany. There is a tiny spot of white fir at the top of Bald Mountain, the Air Force–commissioned report noted; soon it would be interrupted by a new high-tech emplacement of antennas and helipad. Thanks to the land closures, the law required archaeological investigation, which showed the area dotted with petroglyph sites, even a well-preserved nineteenth-century wooden wickiup.

Because the military had to be sure no endangered species were affected, Dreamland became one of the most carefully documented areas in the United States.

It makes me feel good about my country that tanks and nuclear tests are dependent on the cooperation of desert species. At Fort Irwin, California, to the west of Dreamland, military maneuvers are required to stop if soldiers encounter the endangered desert tortoise.

Both the Nellis Range and the Nevada Test Site must have their withdrawal from the public lands regularly renewed, which resulted in an environmental impact statement prepared in 1995–96 for the whole test site. It ran to five fat purple spiral-bound volumes.

Once, the map was blank. Once, the place was real. "One of the most desolate regions upon the face of the earth," First Lt. George Montague Wheeler called it after leading Army Corps of Engineers expeditions through the area in 1869 and again in 1871. It was tough territory, and Wheeler reminded readers of his report that his expedition took place "amid the scenes of disaster of those early emigrant trains who are accredited with having perished in 'death valley.' " He was referring to notorious reports that in 1849 part of the Death Valley Party en route from Utah to California decided to take a shortcut, and camped near Groom and Papoose lakes. Only the intervention of the friendly Paiutes saved them from dying of thirst and starvation.

Unlike earlier expeditions dedicated to science, such as Clarence King's landmark exploration of the 40th parallel a few years earlier, Wheeler's mandate was "reconnaissance": to map the area, survey minerals and mines, and help guide "the selection of such sites as may be of use for future military operations and occupation"—a neat foreshadowing of the later uses of the land.

On his first foray, in 1869, Wheeler and party camped at a place he called Summit Springs, between Pahranagat and the Jumbled Hills, not far from the heights from which the Interceptors would later survey Dreamland. In 1871, on his second expedition, escorted by a detachment of the Third U.S. Cavalry, Wheeler encountered the Paiutes, whom he described as "raising corn, melons and squashes" and harvesting wild grapes. Of this people, who had rescued the California-bound travelers, he added, "Virtue is almost unknown among them and syphilitic diseases very common."

Wheeler's photographer, called "the Shadowcatcher" by the Paiutes, was the renowned Timothy O'Sullivan, who not only left us with the first, lasting images of such wonders of the West as Canyon de Chelly but as one of Mathew Brady's photographers had recorded dead sharpshooters in Devil's Den at Gettysburg. O'Sullivan's photographs of Wheeler's party show men who look even harder than those better-known Civil War veterans. Hard-bitten, resigned, they were as used to fear in this landscape as in battle. Their faces are darkened by the sun above full beards and long sleeves.

On July 23, 1871, Wheeler's geologist, a friend of O'Sullivan's named G. K. Gilbert, visited Groom Mine, and the party's report described it as "one vast deposit of galena," a low-grade ore, mostly lead containing some silver, zinc, and copper. An advance party had been sent to the west, toward Death Valley proper, but that very night their guide disappeared—apparently deserted them—and they very nearly died. The men were down to their last mouthfuls of water before coming on a green spot they immediately named Last Chance Springs. A second guide vanished and Wheeler wrote, "His fate, so far, is uncertain; that of any one to have followed him in the particular direction he was taking when last seen would have been CERTAIN death."

After leaving his campsite at Naquinta Springs, Wheeler headed west, trying to link up with the side party. The hills gradually opened up a prospect of Death Valley that, Wheeler wrote, "met our eyes in strange and gloomy vibrations through the superheated atmosphere."

The same sense of foreboding landscape—more desert hallucination and nightmare of thirst than American dream—emerged in the maps Wheeler's expedition produced as well as in O'Sullivan's photographs. Before Wheeler, maps depicted the interior of Nevada as a great blank space, hostile, rough, forbidding. His cartographer, Louis Nell, filled it with the caterpillared hatchings of hills and lava flats, the warty peaks and scars of passes—a geological history of calamitous events. Fuzzy hatchings—whisker lines—mark the dry lakes. "Groom Mining District" and other mining districts appear as neat boxes overlaying the scarred landscape. The Black Metal Mine a mile south of Groom is shown, along with the road to Indian Springs, now closed off, and another back to the east and Hiko. Like square bandages on a tortured face, the upright lines of the mining districts—the only political markings on the map save roads and tiny towns—reveal civilian settlement that is no more than stopgap.

Today the sense of foreboding, the terror, comes across as stark beauty. Photography critics would later note that these government-financed documentary photographs, with their deadpan alien landscapes, resembled those taken by lunar or Mars landing probes.

———

Wheeler had noted that while there was wood and water in abundance, Groom Mine was not being worked. In September 1872 claims were filed by J. B. Osborne and partners in the White Lake and Conception Lode and British capital was invested to begin production.

The area was not called Groom until the end of World War II, when a geologist named Fred Humphrey surveyed it for the Nevada State Bureau of Mines. Previously it had been called the Naquinta Mountains or Tequima Range. Humphrey found the whole area "imperfectly mapped," and took the range's name from the Groom Mine, after a man named Bob Groom, who was on his way to Oregon when one day in 1864 he came across a promising chunk of ore. Groom never got rich from the claim and never mined it commercially, but he lent the mountains and the lake nearby his name.

Not until a family named Sheahan took ownership in the 1880s did any successful production begin; the Sheahans would keep the mine open through war and thin times, to the present day. Silver was the first goal of the miners, but lead became the mine's main product. More silver was found in the nearby Pahranagat district, inspiring the 1866 Nevada legislature to create Lincoln County. Silver had driven the creation of the state of Nevada and would fuel its subsequent booms. At Dreamland, of course, the goal would be to find "silver bullet" weapons.

Fred Humphrey's photos from the fall of 1944 show a quiet desert landscape, the lake smooth and empty except for shells from wartime gunnery practice. Humphrey mapped the faults—the graben, in geological terms—that served to concentrate lead and silver. His published report includes painstaking orange and blue foldout maps of the rock formations, shale battling limestone, jagged and zigzaggy as an abstract painting. Two huge masses of distinct rocks had pressed together. The result was like Dreamland itself: Where the strata overlapped—on the faults—substances became compressed and concentrated. To understand the strange dark history of the place, I had to explore the cultures in which it was born.

6. "The Great Atomic Power"

Like Paris with its arrondissements, or Chicago with its political wards, the Nevada Test Site is divided into numbered areas. But the numbers seem scattered at random on the map of the mostly rectilinear areas. From one perspective, the outline of the test site looks like a squared-off bird, a canyon wren, say, with its beak at the northwest formed by Pahute Mesa, Area 20, and its stubby tail, to the southeast, by Area 23 and the site's company town, Mercury.

When the grid of the numbered areas dropped like a net over the rough geological and topographical charts in the 1950s, Groom Lake became Area 51, unfolding like a wing to the northeast. Dreamland was not just an offshoot of the NTS but, like Godzilla and a hundred other science-fiction monsters, the incidental product of nuclear testing, a mutation of Cold War thinking.

I was trying to make sense of the map of the site, conscious that soon I would be sitting in the most powerful seat of the century, the big Naugahyde chair in the Command Post of the test site from which an entire nuclear arsenal had been detonated.

———

I would circumnavigate the whole of the Nevada Test Site and the Nellis Range of which Dreamland was the center or, as I often thought of it, the critical core of the bomb. I stopped at Indian Springs, the little airfield from which B-29s and B-50s had taken off to drop the first test bombs in the early fifties. I passed the legal whorehouses of Nye County, lonely trailers surrounded by pickup trucks, gas pumps, and red lights out by the highway—big red dome lights visible from a couple of miles down the road, the sort you might see atop a fire engine. "It's the only place in the world where you can fill your tank, change your oil, and get a blow job all in one stop," Derek joked.

Derek was the man who guided me through the test site. He worked for the Department of Energy and took people through the site for a living, spending whole days driving across Jackass Flats and Yucca Lake and Paiute Mesa.

Derek and I drove up from Las Vegas in one of the earliest snowfalls on record. Eighteen-wheelers had slid into the median and a pickup truck was turned over not far from a billboard offering tax-free cigarettes on sale at the Indian reservation. "I've never seen it like this," Derek said as the snow swirled thicker.

We turned off at the entrance to the test site, rumbling across a cattle guard. What we saw first was "the Pen"—the chain-link-fenced yard that had regularly been used to hold anti-nuke protesters, women on one side of a divide, men on the other.

The sign above the main gate that reads WELCOME TO THE NEVADA TEST SITE AND ENVIRONMENTAL RESEARCH PARK invariably elicits snickers. I clutched my map as Derek drove. It marked the territories of the nuclear death's-head, the varieties of nuclear obsession and fantasy and fear. Here the bizarre nuclear ramjet engine for aircraft had been constructed; here were conducted tests of weapons accidents and waste spills, the ones code-named Broken Spear and Bent Arrow. Here "Grable," the nuclear cannon, was fired. JFK visited the nuclear rocket Nerva, on which once rode our hope for trips to the planets. Surrounded by his entourage, he stood in sunglasses, looking up at the tortuous pipes of the test stand. At the top of the map was the amazingly named Climax Spent Fuel Facility; to the left, and west, the Yucca Mountain project, for planned storage of nuclear wastes into the next several millennia.

Derek, I learned, was a child of the Blitz. He had been taken from London to the country when the Germans began the first exhibition of airpower as terror. His father had been in North Africa with Monty, and during six-hour cease-fires he and his fellows had played soccer with the Germans, then gone back to trying to kill each other.

Derek flew helicopters in Vietnam. He had "taken two armor-piercing in the stomach" and swore the Vietcong paid their troops a twenty-five-dollar bonus for every chopper pilot they took out. Before he came to Las Vegas and the test site, he had worked for the DOE in Colorado, and a discussion of Denver Bronco star quarterback John Elway was one of the few things that brought a smile to Derek's face.

———

The base camp at Mercury provided an inventory of government architecture, from Nissen huts to pastel cinder-block apartments. A sign in the cafeteria advertised an upcoming bowling tournament.

The road through the site runs from the highway turnoff at Mercury and, if you could cross the Ridge, on to S-4—Papoose Lake, putative site of the saucer base. Along the road were old signs warning about security and safety, their stridency muted by wind and sun, which had brought the grain of the plywood back up through the paint.

At Frenchman Flat, to the south end of the site, I stood on ground zero—on many ground zeroes, actually. Most of the first blasts were set off in the air or from towers and balloons in Jackass Flats. But nothing now was hot; the radiation had long since faded and in places some of the top layer of soil had been removed.

We pulled up to a set of test structures constructed at Frenchman Flat for the 1957 explosion called Priscilla. It was a virtual sculpture garden of shapes: an underground garage entrance, built to test garages as fall-out shelters; concrete dome shelters, spheres with just their bald tops protruding from the earth; the remnants of a railroad bridge trestle and a safe contributed by the Mosler company from whose concrete sides the rebar was pulled back like the bones of a cooked trout; the twisted forms of airplane hangars, reddened with rust, like an Anthony Caro sculpture executed in fast-rusting Cor-ten steel. A series of concrete boxes used to test blast resistance and known as the motel or the sugar loaves suggested a Donald Judd sculpture.

The artifacts of testing looked like art. But it also worked the other way: These shapes had inspired Michael Heizer, Robert Smithson, and other ambitious creators of sixties-era "earth art" to leave behind sculpture as monumental as the Ozymandian works of ancient civilizations. In our own time, no one had come closer to putting timeless marks on the face of the planet than the boys at the test site.

Looking at the expressionism of twisted girder and the minimalism of repeated cubes, I suddenly understood how much the reductivist endgame of modern art had in common with the no-win endgame of nuclear warfare. Was anything more abstract than mutual assured destruction? Was it an accident that the end of modernism and the end of the Cold War came almost simultaneously?

We paused at the Sedan Crater, in Area 10, recently added to the National Register of Historic Places.

It is 325 feet deep and 1,280 feet wide and was created by a hydrogen blast on July 6, 1962, the part of Edward Teller's Plowshare program aimed at devising peaceful uses for nuclear explosives. A few tumbleweeds had gathered in its bottom like dust bunnies in an ill-kept apartment. Say a Third World dictator whose country owns a major canal balks at renewing the treaty lease. Well, Teller figured, you just light up a few nukes and dig a new one in the country next door.

Sedan lifted eleven million tons of earth into the air in a blossoming explosion that took on the shape of a great shrub above the desert. It jolted the ground with the force of an earthquake registering 4.7 on the Richter scale. Apollo astronauts in training used the crater to simulate one on the moon.

We stopped by Doomtown, in Area 1, where a couple of houses still stood from the 1955 Apple II blast: little bits of Levittown in the desert, stocked with mannequins from the JCPenney department store and canned and frozen foods flown in from Chicago the night before the blast.

I felt as if Derek were a real estate agent and I a prospective buyer, checking out the dry, gray plywood floors. I stood in one of the living rooms for a while, then walked around the place, as if considering the landscaping. I noticed that the chimney had been twisted on its axis so that bricks protruded a couple of inches.

To the north is another little village of test structures, called Japan Town, where realistic Japanese buildings were exposed to fallout in order to compare the results with the effects of the blasts at Hiroshima and Nagasaki.

From a long way off, behind the Control Point, I could see the Device Assembly Facility, the DAF, which had been built for assembling nukes at a cost of $100 million just before testing stopped. The DAF looks like a huge, long bunker or a giant surfacing submarine. Inside are special rooms whose roofs are slung on wire cables so they will collapse and trap blast and radiation in case of an accidental explosion. The DAF was surrounded by watchtowers, wired fences, video cameras, and high-tech radar sensors on poles. It cried out to be included in a movie and it occurred to me that we taxpayers ought to recover our expenditures by renting it out to Hollywood.

————

We passed "News Nob," where, by the early fifties, any reporter worth his typewriter, any broadcaster worth his mike, had to see an A-bomb for himself. Congressmen, aides, and top government officials were brought here as well, and for the St. Pat's blast of 1953 the revelry was at its peak. The officials of the test site hosted a group of journalists who produced upbeat stories in publications from *The New York Times* to *National Geographic*.

We dropped by Command Post One, the blockhouse control center. Out front were guards in camou. They opened the doors and flipped on the lights for us. It was cold and quiet inside. No explosions had been set off for a year and a half, and the building had the slightly musty smell of a vacation house left closed and vacant for a long time. Inside the control room the thick wooden tables and consoles turned out to be Formica, and chipped at that. I sat at last in the chair from which the big booms had been set off. It seemed cheap, the size pitifully pompous, like the chair of a minor county functionary full of his own importance.

Everything in the room felt years out of date, almost seedy, more like the furnishings of a government health clinic than the powerful high-tech control center seen in old newsreels. The telephones had Lucite cube buttons you punched to choose a line. Next to one of the buttons I saw the designation "dremland" (*sic*). It was a line used to coordinate test operations with the tower at Groom. I surreptitiously jotted down the number and I imagined calling it from all sorts of places around the world, staying in touch with the tower in Dreamland.

————

We stopped for lunch back at Mercury. In the afternoon, we returned to the distant northern part of the test site. Here, Derek said, coyotes and deer roamed. They had lived so far from human contact for so long they would often come right up to you. "They have no fear at all," he said. "It's as if time had stopped." The talk turned to the other side of the

Ridge, and the parts of the site we could not visit. "Area 51?" Derek said. "I'm probably the only one out here who knows what they are really doing over there."

Derek looked at me, gauging my reaction. I didn't dare to ask: "So what is it then?" Because the answer would be either the serious "Well, of course I can't tell you" or the facetious, clichéd joke: "Well, I could tell you but then I would have to kill you."

I looked over toward Gate 700, and it occurred to me that this might be the closest I would get to the heart of Dreamland, Groom Lake, and certainly, in physical distance, to its mysterious sibling, Papoose Lake.

I just let the question, as Henry James would say, "hang in the air."

———

A few days later I met one of the men who had helped build the road I saw running off through Gate 700, connecting the NTS to Dreamland. At noon one hot day I drove through a quiet suburb of Las Vegas. It was empty and silent, neat little houses on neat little lots. Modified ranch with a slight Mexican accent. Stucco. Lots of ironwork. Pastels. Neatly clipped lawns.

Joe Bacco was sitting on his porch. He had worked for years as a maintenance man, fixing roads and other facilities at the nuclear test site and in Area 51. He wore on his identification the number "8," which allowed him to cross the border into Area 51. We talked in the dining room, under the eyes of a Madonna on the wall.

I met Bacco at a hearing on the future of the test site. After the high-pitched Greenpeacers and the Shoshone nation reps and the man who said he had worked with plutonium daily with no ill effects had spoken, Bacco got his turn.

Joe Bacco sweats constantly now. There is a perpetual thin sheen over his body, as if he were in a New Orleans August instead of the dry Nevada desert. His eyes, always partly closed, as if swollen, glisten like his body. Bacco takes showers every few hours.

In 1970 an underground explosion called Baneberry leaked, sending a towering cloud, mushroomlike in shape and size, above the flats and cracking the ground like an earthquake. The fissures were two or three feet wide in some places and made the roads into Area 12, site of the blast, impassable.

The camp at Area 12, where some nine hundred workers lived in trailers and, sometimes, tents, was swept with fallout. Three hundred were found to be contaminated with radiation. The NTS authorities panicked. The radiation release was a PR nightmare; sabotage was suspected. The authorities immediately sent Bacco and a crew of other workers to patch the road. The members of his crew are almost all dead now, he tells me. "It was hotter than a motherfucker," he said, referring to radiation.

"The foreman was Herschel Baker, and there was Charlie Archulet, who's dead now." He lists the names of his other crewmates. "We had to put chains on the four-by-four."

It was snowing heavily that day yet sparks flew from Bacco's long johns. It was so hot, the workers' safety badges were quickly overwhelmed with radiation. "There was electricity all over my body," he told me. "Red and green sparks.

"Later I was paralyzed, and I was passing blood for six or seven months."

It was an account full of primal fear, as much from what he had seen as experienced.

He talked of men who had fallen asleep in trailers before the blast and been killed. He had hauled out bodies. Beside the baseball diamond in Mercury, workers had burned dead cattle and drums of waste, incinerated the badges that recorded how much radiation the workers had received, "to hide the evidence."

One of the men contaminated by Baneberry, the supervisor, Harley Roberts, fought the AEC and later DOE, and helped win rights and recognition for the workers. Baker and others in the crew developed leukemia within two years of the shot. In 1972 Roberts and a worker named William Nunamaker filed suit for some eight million dollars against the NTS, charging negligence. The case lingered on for ten years as the court kept postponing judgment. But by 1974, Harley Roberts was dead.

Bacco's requests for benefits had been denied. Both his old employer REECO, Reynolds Electric, the largest contractor at the site, and the Department of Energy claimed to have no record of his employment, even though he had his work identification card. "They thought, This is a sucker, we use him," Bacco said. "I was a guinea pig."

Where had I heard stories before of employment records being made to disappear? In Lazar's tale, of course.

At the hearing, Bacco told his story with a practiced rhythm. He explained how the Department of Energy had tried to settle with him.

"The lawyer offered me twenty thousand. I told him a big bad word.

What I wanted was my job back. I talked to the doctor. All I said was, do me a favor, when I die give my body to research.

" 'Well, Joe,' the doctor said, 'you ought to feel lucky you're still living. Just keep taking those showers.' "

———

The lady from DOE shook her head sadly. This sort of thing was all supposed to be in the past for the department. Yes, mistakes had been made, but a new page had been turned.

The original creators of the test site were motivated by nothing less than a desperate need to save the planet. A few thousand acres of land, a few hundred lives, were necessary casualties. They were driven with all the intensity of scientists in fifties sci-fi movies, rushing to come up with a weapon to defeat mutant giant ants or invading saucers. But that was in the past. The lady from DOE explained that with testing stopped, the department was looking for new uses for the test site: A solar energy farm was being considered.

The test site tours were at once part of the new attitude and a revival of the proud tradition of News Nob, where Walter Cronkite, Bob Considine, Dave Garroway, John Cameron Swayze, and others were courted as they reported on the Bomb. DOE was trumpeting its new openness, making available old records, pledging never again to expose soldiers and downwind civilians to radiation.

Derek and I did not discuss the way that the bombs exploded at the test site had affected Dreamland.

Among the newly opened records were documents showing that Dreamland itself had been a victim of fallout and of nuclear blasts, even after U-2 testing began there. Work on the U-2 and later the Blackbirds would be placed at the mercy of the needs of nuclear testing. Even the crews and pilots at Groom Lake were in danger. Kelly Johnson had been concerned from the beginning about the dangers of fallout and, sure enough, the work at Groom would frequently be interrupted with warnings or evacuations whenever testing took place. The authorities debated which tests at Groom Lake, if any, would justify delaying a nuclear test.

The first part of Operation Plumbob was called Project 57, conceived to ensure that a nuclear weapon damaged or dropped in an accident or otherwise broken open would not detonate—even if some of its conventional explosives went off.

The test took place just seven miles from the main base at Groom, in the Groom Lake Valley, near the mine. A ten-by-sixteen-mile block of land surrounding the planned location was added to the test site and designated Area 13.

No one involved with Project 57 seems to have had much of a contingency plan if the bomb wiped out the U-2 program already under way at the lake, not to mention the mine and its operators. Later it occurred to the people in charge that, with the base at Groom growing, this was not a good thing. So in the 1980s the government spent twenty-one million dollars to have the land scraped and the toxic portions removed, a process clearly visible in spy satellite shots.

In June 1957, training for the U-2 pilots was moved to Texas, probably because of the bomb tests. Soon afterward, Project 57 began with a huge blast called Hood detonated from a balloon fifteen hundred feet over Area 9, about fourteen miles southwest of Groom Lake. At seventy-four kilotons it was the most powerful airburst ever set off within the continental United States. There was no public announcement. Fallout descended on Groom Lake, and the concussion shattered windows in the mess hall and a barracks and buckled the doors of two metal buildings.

During the tests, the crews at the new base were regularly warned and evacuated. They were unaware that they were part of a long tradition and that other neighbors of the test site had not been so lucky.

———

During these years, a man named Bob Sheahan assembled a unique photo album of Dreamland. The mushroom clouds rising from the spots I had visited at Frenchman Flat and Yucca Flat were visible from his home at Groom Mine, on a ridge about forty miles from ground zero. He took dozens of pictures of the blasts, a whole catalog of mushrooms—twenty, thirty Hiroshimas, seen from the edge of what was to become Dreamland.

Bob Sheahan had grown up around Groom Mine, with its cluster of work buildings and adjacent cabins. The mine has been in his family since 1885 and his father, Dan, ran it now. Bob was thirty-one, a former engineering student at the University of Nevada, when one day in early 1951 a polite, well-dressed man from the Atomic Energy Commission came calling. There would be atomic blasts, he warned them, at the new proving ground about thirty miles to the southwest, and some radioactive fallout might drift over the mountains. It would head northeast toward them, crossing the Groom Range at Coyote Gap, near the site of what would become the town of Rachel, with its little monitoring site on the town square. The AEC man gave the Sheahans a Geiger counter and taught them how to use it. He left flat sticky plates to catch fallout for later testing. He set up a radio.

Dan Sheahan had the Atomic Energy Commission boys sign his guest book. "We're all family," he said.

The first shot, on February 2, 1951, broke the Sheahans' front door and cracked several windows. Others quickly followed. With the Korean War turning ugly, research into tactical nuclear weapons was pushing ahead hard.

Soon the Sheahans began to see signs of the fallout. Bits of metal big enough to pick up with a magnet, all that was left of the vaporized steel towers, fell out of the sky. The Geiger counter showed the metal was hot.

Strange white spots about the size of a silver dollar began to appear on the backs of cattle and horses. These, the AEC man would tell them, were called beta burns. One day Sheahan saw an object on the ground, and when he got close he found it was a dead deer, marked with the same white spots as the cattle. He noticed something else strange: There were very few rabbits. Usually, the desert was full of them—you would mount any rise and startle one—but now he hardly saw any.

The first series of shots came in rapid succession. They were part of the series called Upshot Knothole. But the fallout from the series called Operation Buster Jangle was worse. These were run mostly by the Army, which set up a whole tent town at the proving ground called Camp Desert Rock and exposed tanks and troops and all sorts of equipment to the edges of the blast. In one test, the Army tried to determine the effects of an atomic blast on uniforms at varying distances from ground zero. Miniature uniforms complete with zippers, snaps, and toggles were custom sewn to fit each of 111 white Chester hogs. The pig was chosen because, flattering to our species or not, its muscle and fat distribution most nearly resemble those of a human being.

Most of the pigs, each in its specially tailored little pig uniform, ended up barbecued alive, and there must have been a smell of roasting pork that might not have been entirely repulsive. The test was jokingly called "The Charge of the Swine Brigade." But the troops too were being exposed—far more than many knew—to radiation.

On May 5, 1952, soldiers came to warn the Sheahans of an impending very "dirty shot" and suggested they evacuate. Dan and Bob Sheahan stayed; the rest of the family went to Las Vegas. The next day, a blast went off that broke windows and ripped sheet metal from the buildings.

Worst of them all was the ninth shot in the series, code-named Harry, on March 24, 1953. It irradiated some four thousand sheep being herded through Coyote Gap. Within a few days they would all die.

The fallout from the Harry blast traveled as far as St. George, Utah— to the northeast—with deadly effect. Years later the trail of cancers it left among the "downwinders" became the subject of lawsuits. By the nineties, however, it was clear that most of the American population had been downwinders. A report credited the blasts with causing some seventy thousand cases of thyroid cancer alone.

After "Dirty Harry," cattle drinking from Papoose Lake died, but the Sheahans still felt the AEC was taking care of them. They once made a trip to the office at the test site. An officer forthrightly explained to them that the shots were set off when the winds blew toward Groom, to avoid sending the fallout toward Las Vegas.

Once, the soldiers came to check the Sheahans' water hole. They took samples and the sergeant assured them it was fine. Then one of the enlisted men asked if he could have a cup of water. "Can't you wait until we get back to camp, soldier?" his commander gruffly interrupted. When the men realized the implications of their exchange, both became silent and embarrassed.

During all this time, Dan and Bob Sheahan had to halt operations at the mine, sometimes for two weeks at a time, because of the tests. Nor was the mine safe from conventional weapons. It was still part of the gunnery and bombing range, and in 1954, an overeager trainee strafed the mine buildings, presumably mistaking them for one of the target buildings on the range.

Finally Dan Sheahan discovered that his wife, Martha, had cancer. He would eventually sue the AEC, but the Sheahans held on to their land and mine, passing it to the next generation, Pat and Bob, and worked out an uneasy truce with the Air Force. But Bob never showed off his photographs, and into the nineties he was afraid to talk at all about the mine lest the Air Force make his life difficult.

By the seventies, Martha Sheahan had wondered how the military could say they were defending freedom at the base while trampling on the freedoms of those on its edge. But after the guards showed up at the mine in 1984, the Sheahans fell silent. At least some of the family were given security clearances, and when I talked to them in the mid-nineties other family members were still unwilling to criticize what the government had done. "They take care of us," one family member said of the Air Force. He refused to talk. He didn't want to be identified.

———

The dirty blasts of the early fifties baptized Groom and Papoose lakes in radiation. And the base that would grow up there, like a gigantic sci-fi mutant, would share the ethos of emergency, justifying the pollution of the "unpopulated" areas around it.

In its own irrepressible way, Las Vegas seized on the proximity of the test site in a more festive manner. The bright boomerangs and bubbles of neon on the Strip arrived just about the same time the flying saucers did. In honor of the destruction of Doomtown, the suburban town built in 1955 for the Apple II explosion, one Vegas hotel filled its swimming pool with mushrooms. Parties assembled to watch the blasts from convenient high spots. There were picnics on Mount Charleston, halfway up to Mercury, a future site of Interceptor expeditions. Even weddings were scheduled to coincide with nuclear tests: honeymoon in Las Vegas! Did the earth move for you too, dear? The mushroom cloud became another party theme, like the themes of the Old West, the Middle East, Ancient Rome, invoked as keynotes for decor at the Frontier, the Sands, or Caesars Palace. The Flamingo served an Atomic Cocktail—vodka, brandy, schnapps, and a touch of sherry. Gigi, its top hairdresser, arranged wire to produce an Atomic Hairdo. In May 1957 the Sands held a Miss Atomic Bomb contest in which the competing beauties appeared with the iconic mushroom cloud, modeled in cotton, glued to their silvery swimsuits.

———

Las Vegas is hardly typical of the United States, but for a time the whole country shared in the eagerness to embrace the atom. The historian Paul Boyer calls it the search for the silver lining to the mushroom cloud. There was an effort to downplay the effects of fallout and blast—it was actually proposed that a good wide-brimmed hat could offer a lot of protection—and civil defense drills became a common activity for schoolchildren. The stylized logo of the atom, with its zippy futuristic orbiting electrons, was soon joined by the three triangles on yellow of the fallout shelter as nuclear age icons. Disney published a children's book called Our Friend the Atom, and the Boy Scouts added an atomic energy merit badge to their sashes. But beneath the cheery atom culture—so well documented in the 1982 film The Atomic Cafe—was a deeper and frequently denied fear. The atomic bomb shook heartland America to the core.

While Las Vegas was dancing to "The Atomic Bounce," country-and-western music struggled to deal with the darker fears of the bomb. As I drove the fringes of Dreamland, I often played tapes of music from the fifties. One song in particular seemed to sum up poignantly middle America's effort to deal with the shadow of the mushroom cloud. "The Great Atomic Power," by Ira and Charlie Louvin and Buddy Bain, documented the bomb's impact on the nation:

> Do you fear this man's invention that they call atomic power?
> Are we all in great confusion?
> Do we know the time or hour?
> When a terrible explosion may rain down upon our land,
> Leaving horrible destruction,
> Blotting out the works of man.
>
> Are you ready for that great atomic power?
> Will you rise and meet your savior in the air?
> Will you shout or will you cry
> when the fire rains from on high?
> Are you ready for that great atomic power?

The Louvins' song belonged to a tradition of songs beginning with "Atomic Bomb," penned by the sleepless Fred Kirby the very night the first bomb was dropped on Japan, August 7, 1945. Recorded by many groups, "Atomic Bomb" was joined by such numbers as "The Hell Bomb," "Jesus Hits Like an Atom Bomb," and similar songs, which were big hits in the late forties and early fifties.

Like others in the genre, "The Great Atomic Power" was a conflation of Bible and Cold War, a rendition of the apocalypse as nuclear holocaust. The bomb's coming was the Second Coming and you'd better be ready, better turn to Jesus for salvation. The song was a desperate, even heroic, effort to graft the impact of the bomb onto fundamentalist Christian theology, to force the terrible new knowledge into the net of traditional teaching and, grotesquely, deform it. Here was Jesus as the ultimate version of Strategic Air Command—"He will be your shield and sword"— right off the logos painted on the noses of B-36s and B-47s.

"When the mushroom of destruction falls in all its fury great, God will surely save his children from that awful awful fate."

SAC's Gen. Curt LeMay, however, wasn't waiting for God; his plan was to hit the Russians with everything he had before they could light up the skies over New York and Washington, over Dallas or Nashville, or over Omaha, home of SAC, seat of the religion of airpower.

7. Victory Through Airpower

Embracing the Nevada Test Site and looming over Las Vegas on the map, the Nellis Air Force Base wrapped Dreamland in the ideology of airpower.

The huge bright tank of jet fuel at the entrance to the base read GLOBAL POWER FOR AMERICA. Emblazoned with the shield and sword of the Air Combat Command, the tank shimmered in the heat just up the road from the pawnshops and watering holes (SNAFU Lounge) that have sprung up on the verges of the base.

Dreamland is part of Nellis's vast bombing ranges, but Nellis is best known as the home of the Air Force's Red Flag training games, the equivalent of the Navy's famed Top Gun school. During Red Flag exercises the sky for hundreds of miles around the base is filled with aircraft—twenty-two thousand sorties are flown a year.

I stopped by the edge of Nellis's runway one afternoon during a Red Flag to watch the airplanes returning. A half dozen or so cars and trucks had gathered, with people lounging in the driver's seats or sitting lazily on the hoods. Their expressions bore the patient, purposeless air of fishermen.

F-15s and F-16s came home in pairs, each touching down with a little puff of smoke as its tires hit the pavement. A big AWACS plane, a huge hump of an antenna on its back, came in over our heads, and a helicopter drifted past, creating a little crater of dust.

For Red Flag, the planners at Nellis are constantly creating "notional countries," imagined allies and aggressor nations that play out the scenarios of conflict, drawing fictional nations on the map of the area around Dreamland. In one Red Flag scenario, for instance, a friendly little country named Cavalier is menaced by aggressive Sirocco.

In the absence of war, or rather in the Cold War the military has fought in the last half century, the game is the thing: Witness the constant playing of war games, from the high level of Herman Kahn, the doomsday theorist of nuclear holocausts, to those of Top Gun and Red Flag.

Geopolitical scenarios played out by Pentagon planners are popularized in the military technothrillers of Tom Clancy, Dale Brown, Harry Coyle, and others. Dreamland crops up in them repeatedly. Clancy makes reference to flying saucer lore when he has a character joke about "the Frisbees of Dreamland." Brown, who flew in Red Flags during his Air Force days, describes a fictional High Tech Aerospace Weapons Center, HAWC, at Groom Lake in his novels *Sky Masters* and *Flight of the Old Dog,* "a secret U.S. Air Force research facility in Dreamland that conducts flight-test experiments on new and modified aircraft and new weapon systems."

Sky Masters—dedicated to Curtis LeMay, "the Iron Eagle"—describes the testing of fuel air bombs, a real technology in which a huge cloud of gasoline vapor is ignited, producing a shock wave that crushes troops on the ground—a miniature Tokyo firestorm.

For Nellis, Dreamland is always "the Box." Military Operating Areas, forbidden to civilian air traffic, show up on aviation maps, marked with the acronym MOA—an unintentional irony, since the moa, a now extinct bird from New Zealand, was flightless. They are given names like Talon and Cheyenne.[1]

Pilots take the Box very seriously—because their commanders do. It is the most restricted MOA, off-limits even to the military pilots, at all altitudes and all times.

At the beginning of every Red Flag session crews spend several hours and one two-hour sortie being oriented to the various Nellis ranges, memorizing landmarks so that in the heat of "battle" they do not stray into the Box. Even crossing buffer zones around R-4808 (airspaces R-4807, R-4806, R-4809) results in the crews being given a slap on the wrist, but it happens frequently.

A former Red Flag player explained to me, "If a pilot accidentally strayed into the area, the day's exercises would immediately terminate and the offending aircraft would be ordered to land at an isolated area on the east side of Nellis AFB. Intelligence officers would confiscate the radar film, detain the crew, search everything in the cockpit, and then conduct a lengthy interview to determine why there was an overflight. Overflying R-4808 is cause for very heavy penalties, including an automatic Article 15 [administrative reprimand, which for officers is the kiss of death], demotion, and loss of pay. If the overflight was intentional, one could expect a court-martial, a dishonorable discharge, and imprisonment."

Nellis was created as part of the network of bases built up in the vast West in anticipation of World War II—the Las Vegas Army Air Corps Gunnery School, established in January 1941. By June, the school was graduating four thousand students every five or six weeks. A number of auxiliary runways were built in the huge expanse of the range, including two five-thousand-foot runways laid out in a cross tilted to the northwest on the edge of Groom Lake. Soon the lake bed was littered with .30- and .50-caliber shells.

After the war, Nellis served as a major mustering-out point for airmen and soldiers. It was closed down in 1947 but reactivated two years later, in time to become the main fighter-pilot training center during the Korean War and, eventually, the temple of fighter tactics and esprit. It would become the home of the Thunderbirds, the Air Force aerobatic team.

Today, in the dry cleaners and pizza joints surrounding Nellis, proud entrepreneurs display signed Thunderbird photos, Thunderbird banners, Thunderbird plaques. The Air Force aerobatics team figures here something like the football teams in other cities. There are weekly tours of the Thunderbird hangar, led by a disarming PR man. On the tour I joined, the crowd was largely oldsters. "Aren't they handsome?" said one woman, looking at the photographs of the pilots on the wall. In an auditorium, the guide sketched Thunderbird history and glory in slides and narrative. At the end of the presentation, questions were entertained. Immediately came the impertinent inquiry from a retiree at the back of the room: "Do you have anything to do with this Area 51?" A faint scattered laugh came from the knowing minority.

"That's where we get our pilots," said the PR man, quick-witted. "No, seriously, that's a good one. I wish I knew."

On a ridge above the dry lake called Muroc, one hundred miles north of Los Angeles, some fifty years before I stood on Freedom Ridge, a loose gaggle of men stood shivering in front of small fires. In the hour just before dawn on January 8, 1944, some two dozen engineers and workers of the original Lockheed Skunk Works awaited the first flight of the jet fighter they had produced in only sixty-eight days, the XP-80.

Among them was a man named Wally Bison, who had worked in the Skunk Works from the beginning.

"It was cold, colder than a well," Bison told me years later. "All of a sudden somebody said, 'Here he comes,' and the airplane passed by a couple of hundred feet off the deck in dead silence. Then the jet blast came, a sound we'd never heard before. I was goose pimples from the top of my head to the bottom of my feet." That sound, he said, seemed to come from everywhere and nowhere in particular.

The plane was called *Lulu Belle*, but the guys around the shop had wanted to call it "the Fartin' Fury of Forty-four." These men worked on a relentless schedule; the big legends above the girlie calendars gloomily read, "Our Days Are Numbered."

The formal name of the facilities around Rogers Dry Lake, where Jack Northrop and other aviation pioneers had tested planes in the twenties, was the Muroc Bombing and Gunnery Range. As early as 1933 the Air Corps had established a gunnery range on the lake bed not far from a small settlement established by Clifford and Eve Corum in 1910. When the Corums had applied to set up a sub–post office under their name in the store for the convenience of customers, the authorities replied that the name Corum had already been taken by another office. So they reversed the letters and it became Muroc. The word, with its accidental overtones of Morocco, of the French *mur* and rock, had an appropriately rugged, dry sound.

With the coming of the war, the military arrived in force at Muroc and other dry lakes—China Lake, El Mirage—and chased off the hot-rodders and cyclists. The Navy built a wooden mock-up of a Japanese cruiser, a gray looming practice target, with only the dry lake for waves. They named it the *Muroc Maru*, and for years it floated on the liquid of mirage. Within months of Pearl Harbor there were thousands of men and hundreds of bombers and fighters here and at tens of other new bases springing up throughout the West, safely inland and isolated.

The airplane that flew that day in January 1944 was the second "black" aircraft, the first product of the Lockheed Skunk Works and as secret as the Manhattan Project. Bell Aviation had built the first, the XP-59, and flown it here in 1943. Optimistically named "Aircomet," the XP-59 turned out to be little better than the best prop planes of the day—a lesson from the beginning that black projects could turn out turkeys as well as eagles.

But the project had already begun to display the little signs of camaraderie and conspiratorial clannishness of black projects to come. The Bell crew took the derby hat as their symbol and would fly wearing derbies and gorilla masks, waving cigars as they buzzed hapless fighter trainees, who nearly fell from the sky in shock. Forbidden to acknowledge their work openly, they sported insignia from which the propellers had been removed.

When they went out to Juanita's in Rosamond, and Pancho Barnes's famous bar closer to the base, they wore black derbies and fake mustaches from a Hollywood prop store.

But once the Lockheed jet flew, the XP-59 was doomed. Lockheed's XP-80 proved a durable design. Quickly improved with a more powerful engine, it became the YP-80A, "the Gray Ghost." Eventually some six thousand aircraft based on the type would be produced, a whole family of jets, including the P-80 and F-80 military fighter, the Shooting Star, and the T-33 trainer.

————

That cold January day, the head of the Skunk Works, Kelly Johnson, stood impatiently by the airplane in a long overcoat and knit watch cap. Johnson, then just thirty-three years old and the designer of the Lockheed Electra, the P-38 Lightning fighter, and—for Howard Hughes—the lovely tri-tailed Constellation airliner, had flown to Wright Field in Dayton just the summer before. At one-thirty on the afternoon of June 8, 1943, he had been handed a signed contract to build the airplane that was now complete.

Wally Bison, the Skunk Works veteran, remembered something else. An old man when I talked to him, he had a hard time recalling names, and he kept apologizing. But Bison remembered that Johnson took the whole gang to a restaurant for lunch—and paid for it. Used to bringing his lunch to work in a brown bag, Bison saw this as an act of unprecedented largesse on the part of the penny-pinching Johnson.

When the jet contract arrived, Johnson had to build his own secret team. All of Lockheed's production capacity, all of its engineers and workers, were so pressed to meet normal wartime contracts for the P-38, for B-17 bombers for the Army, sub hunters for the Navy, and Hudson Bombers for the British, that Johnson had to scrounge a staff of twenty-three engineers and about twice as many mechanics, fabricators, and clericals.

Johnson would often fix or construct some part himself. He was proud of the strength he had acquired putting up lath as a teenager. He had been the seventh of nine children from a poor Swedish American family in Michigan and baptized Clarence, taking the name Kelly when his classmates deemed his sometimes violent streak more Irish than Swedish.

Johnson grew up reading Tom Swift and the Rover Boys in the local Carnegie library in Ishpeming, Michigan, a mining town that sent its ore to Carnegie's mills. He built dozens of model planes, and by the time he was twelve he knew he wanted to be an aircraft designer. He put himself through the University of Michigan by washing dishes and putting up the plaster lath, and had been at Lockheed since 1933, when he was hired at a salary of thirty-three dollars a week. His first achievement was pointing out a serious aerodynamic flaw in the prototype Model 10 Electra, on which the firm's entire fortunes rode, and then figuring out how to fix it.

At just thirty-three, Kelly Johnson was already one of the top aircraft designers in the world. He met Amelia Earhart and prepared her Electra for her record flights as well as for her last, fatal journey. He had worked for Howard Hughes on the Constellation airliner that had flown for the first time almost exactly a year before, as the C-69, and already gone into service for the military.

During work on the Constellation, Johnson met often with Hughes, huddling with the billionaire in one of his bungalows. A pilot whose fame had grown from a record round-the-world flight during the 1930s, Hughes tested the plane himself. Once when he took the wheel of the prototype, Johnson and others in the plane were overcome with terror as Hughes attempted to stall the airplane to test its stability. When the airspeed indicator read dead zero, Johnson forced Hughes away from the controls.

Johnson's methods were instinctive and highly practical. He demanded whenever possible that stock parts be used or adapted, and he administered by emotional economy as well: by temper and fear. But he could exhibit flashes of kindness, too.

Bison recalled that "Johnson could be intimidating and brutal, but at our parties he was delightful. When I had to go to Kelly's office, I was in fear, but in the end I was always amazed at his knowledge of the most detailed things." The key to Skunk Works speed was efficient administration. "The main thing was that Johnson cut the paperwork. We drew things upstairs, then walked down and told the mechanic, 'Build the damn thing,' and then you helped him do it."

Johnson and his staff had already been looking ahead to jets. They had proposed a design called the L-133, a stainless steel vision of the future, with a jet engine of Lockheed's own design, a long fuselage, and canards promising a top speed of 650 miles per hour. On May 17, 1943, when Johnson was on a visit to Florida's Eglin Air Force Base, a general took him aside and told him of the XP-59A, that hapless Bell jet, kept secret at Muroc with a fake propeller on its nose. He wanted Johnson to do something better. On the airliner home, his ulcer working overtime, he jotted down the ideas for a jet fighter he thought could be built in six months, something that could be a war winner.

The spaces where the engineers set up their drawing boards and the shops downstairs were cobbled together around a machine shop beside Lockheed's wind tunnel, housed in a leaky addition built from the old wooden crates in which Wright engines had been shipped, the roof made from a rented circus tent.

Working ten-hour days, six days a week, the group put the jet in the air just 143 days after formal signing of the contract. The smell of chemicals seeped into the crude buildings from a factory next door, and an engineer named Irv Culver picked up the phone one day and spoke the immortal words, "Skunk Works." Inspired by Al Capp's Li'l Abner comic strip, Skunk Works is named after the still where a character named Injun Joe brewed up a foul moonshine called Kickapoo Joy Juice. And the question Culver and the others kept being asked but could not answer was "What's Kelly brewing up in there?"

The name was born of secrecy. There was no official designation, so those inside had to dream one up.

———

Johnson continued to call the base Muroc for years after the name was officially changed to Edwards Air Force Base, after Glen Edwards, a test pilot killed in a crash. But by 1955, when the Skunk Works was looking for a place to test the U-2, Edwards was no longer private enough. Like some wild species that needed lots of range or whose environment was changed by the advance of civilization, the engineers who built secret aircraft had to flee farther and farther into the wilderness.

————

The Skunk Works would create the airplanes that made Dreamland necessary, and its legend grew up along with the secret base. It developed a far-flung fan club of buffs as devoted and dogmatic as any group of Roswell believers or saucer conspiracists. It even emerged as a business model, a method to get things done in a lean and mean way, after the management guru Tom Peters wrote approvingly of it. The Skunk Works, the buffs believed, had done nothing less than save the world several times. The U-2 and the Blackbirds had prevented World War III; the Stealth fighter had won the Gulf War.

One day I drove to Burbank to visit one of the Skunk Works' most devoted buffs and see the original site. My guide was a local man named R. C. "Chappy" Czapiewski. Chappy was proof of the power of the legend: He had never worked at the Skunk Works or served in the branches of the military that flew its airplanes, but was simply a citizen who appreciated its achievements and was caught up in its history and lore.

We met in downtown Burbank, which contrary to all of Johnny Carson's jokes struck me as a pleasant place: an inoffensive mall, a new media center, and a series of elegant Modern-style public buildings. The soaring lobby of its city hall was a WPA-era fantasy, painted with romantic murals of thirties aircraft and heroic images of movie cameras— icons of the leading local industries.

A querulous man who spoke with an edge of outrage, Chappy appeared something of a pain in the ass to the local city councilmen. I had to like him right away. Over a Japanese lunch he agreed to take me on a tour. He gave me a yellow-green button that read SOS: SAVE OUR SKUNK WORKS.

He was trying to muster the citizenry of Burbank to save the original Skunk Works buildings from destruction and turn at least one of them into a museum dedicated to the airplanes designed here, from the P-38 to the Stealth fighter.

The organization most opposed to this plan, he told me, was Lockheed itself. The local airport authority coveted the land on which the hangars stood for a planned expansion, and Lockheed had agreed to sell it.

"This was historic," Chappy lamented, "and now it's being forgotten. It was secret, but we—all of us living here—knew what was happening. The U-2, the Blackbird, the Stealth—they won the Cold War. Kids today don't remember the Cold War. They think U-2 is a rock band."[2]

We wandered among the hangars. Crape myrtle trees dotted the avenue in front of them, their pinks and greens virtually the only touch of color across expanses of gray pavement, gray chain-link.

Lockheed had painted the hangars a soft yellow, the yellow of crème brûlée or the yellow rose of Texas. They had chosen the same color out at Helendale in the secret RCS complex.

We could see building number 360 with its complex system of window panels. This, Chappy pointed out, is where they did the F-104— "the Starfighter," "the missile with a man in it"—developed to counter the superiority of the MiG-15s American pilots encountered in Korea. Stubby-winged, with a downward-firing ejection seat, it would be a hot rod, but also a widowmaker, with no more glide in it than a bathtub pushed off a roof.

Here, Kelly and his boys created the U-2, and turned back over to the U.S. government—your tax dollars at work—$2 million of the $26 million he had agreed to accept and a tossed-in half dozen extra airplanes to boot. There, first for the CIA and then for the Strategic Air Command, they built the Blackbirds, the A-12 and the YF-12, then the SR-71, pioneering whole new technologies such as extruding titanium to create a plane that, through the millennium, will be the fastest and highest-flying.

"Beside those hangars," Chappy said, "is where they set up the first Stealth prototype between two tractor-trailer trucks with camou net covering the ends and fired up the engine."

There, the Stealth fighter grew from a mere footnote in a Soviet scientific journal into a black-faceted body. Appearing hacked, chopped, with every corner cut, its shape was the very embodiment of the Skunk Works philosophy, a treatise on the theme of cutting away all excess. "Keep It Simple, Stupid" was the motto—KISS. "Simplificate and add lightness."

At the end of the flight line was the big hangar, where stealth watchers had seen shadowy tarpaulin-covered payloads moving into huge cargo planes—things on flatbed rail cars.

"They were flying something on a big C5A out of here in 1991 and 1992. They would shut the airport down when sensitive cargo was being loaded," Chappy recalled. "They stopped all airport traffic at eleven-thirty on Friday for it."

After a Japanese sub surfaced off the coast of California in 1941, Lockheed put in a desperate call to the Disney studios. Their best artists came in to hide the factory under camouflage. They created an artificial, subscale village atop the factory buildings and airport terminal, a model of the very American way of life they were designing and building airplanes to save, the American dream trumpeted in magazine ads and radio serials. Workers inside turned out P-38 fighters and B-17 bombers, then went home to little bungalows very like the little Monopoly houses above their heads.

From the air, you couldn't tell where the roof stopped and actual houses started. Huge poles held up the netting and camouflage, done in chicken feathers, and the real buildings beneath were painted in similar mottled vegetable shapes. "Someone who worked here," Chappy said, "told me that when it rained the chicken feathers stank to high heaven."

The Skunk Works was the best argument for black projects, but it was always a gamble. There had been failures: the Saturn commercial transport, the F-90 fighter, the weird tail-landing XFV-1 Salmon (named for Herm "Fish" Salmon, the test pilot and only man crazy enough to ever fly the damn thing). The D-21 drone, a secret for decades. Suntan, the liquid hydrogen–powered superplane of the late fifties that cost $2 billion before someone stopped to consider the expense of building bases with cryogenic facilities to keep it fueled. Even the successes were close enough to failures, like the A-12, the Blackbird, which by rights should never have worked and reminded you that this was gambling at very high stakes.

But as I rode around with Chappy, I found the sense of the legend beginning to wilt. I wondered if now, perhaps, the darkness was too great and the gambles no longer paid off.

I kept thinking about a talk I had had with Ben Rich, the last head of the Skunk Works to preside over the Burbank facility, and I remembered his tone. He had written a book, but when he submitted it for review, two chapters had been rejected by the CIA and the Office of Special Investigations (OSI). It still irked him. Why, they had made him lock up his coffee mug, the one that read MACH 3 PLUS with a picture of the SR-71. But that speed had never been officially released to the public; the information was still classified. So each evening the mug went into a safe, and each morning it came out again so Rich could sip his decaf.

Only one other group of people irked Rich as much as the security people: the EPA bureaucrats, who threatened to shut down his program if it didn't comply with regulations. To him, the concern for information leaks and leaks of chemicals into the water table were somehow equivalent. I often wondered if secrecy itself hadn't become a toxin, extremely powerful and useful in controlled amounts, but treacherous and poisonous if misused or overused.

We drove out of the hangar area past a range of Dumpsters. Suddenly, a big plastic bag flew up in front of the car. "UFO!" Chappy cried.

———

The original site of the Skunk Works was now flat and bare, loosely covered with rubble like a site ready for construction. Across the street, bougainvillea climbed concrete walls in front of quiet, well-kept homes. In a cruel irony, the Skunk Works had lived up to its name, pumping PCBs and other pollutants into the surrounding aquifer, and the company had become the object of massive litigation. Lockheed had recently settled with a group of local residents for about $130 million. A huge piping works called a vapor extraction system would pump steam into the ground and pump the toxins out—a grotesque distillery.

To those imbued with the Skunk Works legend, like Chappy, this rubble-strewn field was akin to the fields of Gettysburg or Yorktown or Agincourt and should be preserved in the same ways. But the Skunk Works headed out to Palmdale, closer to their desert test base. What they left was wreckage.

———

Driving away from the dark and gory ground of the original Skunk Works, I passed the Disney complex on the freeway.

While some of his artists were sent to the nearby Lockheed factory to camouflage it, Disney set others to work creating the alluring myth of airpower—one of the great myths that was to propel Dreamland.

In 1943 they were to make the film *Victory Through Air Power*, a powerful piece of propaganda, offering a neat, ideological solution to the muddles of war—technology to keep the distant enemies at bay. Filming began only after ten at night, because there was too much noise during the day from the new P-38s and B-17s taking off from Burbank airport.

Disney made the film at his own cost, so enamored was he with its source: the book of the same name by Count Alexander P. de Seversky. A White Russian émigré who had distinguished himself as a naval aviator in World War I, in which he lost a leg, he came to the United States when the Revolution erupted. He allied himself with Billy Mitchell, the maverick general, and after Mitchell died in 1936 Seversky became the leading exponent of the faith that strategic bombing would be the dominant force in all modern wars.

He was the head of Seversky Aircraft Corporation (the forerunner of Republic Aircraft), and his book, a collection of magazine articles, was a huge bestseller and Book-of-the-Month Club selection. It warned Americans that they could no longer rely on the oceans. They were no longer safe in Kansas City or Chicago. But Seversky held out a promise—if Americans built a massive force of bombers and destroyed the distant cities of its enemies first, we could return to our comfortable isolation.

Seversky laid out the rationale for fighting wars with bombs that would lead to the A-bomb. Soon the poet Randall Jarrell, who served in the Air Corps, would write, "In bombers named after girls / We bombed cities we had learned about in school."

Seversky, called "Sasha," narrated *Victory Through Air Power* in his exotic and authoritative Russian accent. The film blended newsreels of Mitchell and his famous demonstration of how to bomb battleships with cartoon explanations of the development of military aviation. Animated maps explained the present situation, and represented Allied airpower as an eagle, fighting the Japanese octopus, destroying its head as the tentacles slowly released their hold.

James Agee, then film critic for *Time*, found the movie a skillful piece of propaganda, but he noted that it never showed civilians on the ground, never showed the target. And he was disturbed by the climactic battle of the eagle of airpower and the octopus of Japanese aggression and the abstracting of war, with its total absence of images of the victims. It was full, he wrote, of "gay dreams of holocaust."

Richard Schickel has argued in his history of the Disney studios that

Disney liked airpower because it was efficient, clean warfare, in which the corpses are never seen. And airpower especially seemed to have held great appeal to Midwesterners like Disney, Curtis LeMay, and Dwight Eisenhower, who once felt themselves at the greatest remove from foreign influences. Airpower was in an odd way the flip side of the region's traditional isolationism, a way to play world power without sending soldiers overseas. And it seemed cheap, too—in lives and in dollars—a feature that would make it especially attractive in the postwar years.

———

It was not long before Seversky's and Disney's dreams of holocaust would be realized. The B-29, the long-range bomber, was being developed in top secrecy at Boeing in Seattle even as the film was being made.

Curtis LeMay, then training at Muroc, would follow the B-29 from bases in India to China, then the Marianas, from which at last the bombers could effectively reach Japan, Disney's eagle attacking the octopus.

When the first raids aimed at precision failed to strike their intended targets owing to bad weather and bad bombing, LeMay was put in command.

The B-29 was a complement to the A-bomb program. When LeMay took over, crews were training for the A-bomb mission in a godforsaken corner of Utah, near Wendover, living in barracks little better than huts. But the plane had been ineffective in carrying out the high-altitude precision bombing for which it was designed and which was the key tenet of LeMay's airpower theory. So he tried something new—gambling the lives of his crews. He turned to terror bombing: firebombing whole cities. Now his target problem was simpler: find the areas of cities that were the oldest and had the largest proportion of wooden buildings. The first target was Tokyo's Shitamichi district.

Stripping the bombers of most of their guns and sending them in low and at night, on March 9, 1945, LeMay dispatched 334 bombers from bases in the Marianas, each carrying about seven tons of incendiary bombs.

The bombs burned more than the sixteen square miles targeted and killed between 80,000 and 100,000 people. In no other six-hour period of human history had so many people lost their lives. The firestorm was so powerful it sent updrafts that tossed the bombers about as their crews breathed the sickening smoke of burning houses and flesh.

Survivors reported that from the ground the bombers silhouetted against the sky sometimes looked like the black blades of knives and sometimes, when the flames lit them from below, like silver moths trapped in the amber reflections. The bombs themselves seemed to fall like a liquid silver rain rather than a series of solid, deadly objects.

Women fleeing, carrying babies on their backs, continued walking, seemingly unconscious that the bundles had burst into flame. Bodies twisted and turned into the pumice of Pompeiian victims. Those who dropped into canals or pools seeking refuge boiled to death.

The fires died down fairly quickly, and processions of silent refugees moved under moonlight amid the burning ruins. One man paused to light a cigar at a still burning telephone pole.

Time magazine called the raid "a dream come true." It showed that "properly kindled, Japanese cities will burn like autumn leaves."

Approximately as many people died in this, the first great triumph of airpower, as did in the Hiroshima and Nagasaki atomic bombings combined. The step to the atomic bomb was now only a technical one.

———

The dominance of airpower was ratified in 1947 by the establishment of the Air Force as a separate branch of the military, equivalent to the Army and Navy. The same year brought the creation of the Central Intelligence Agency, the declaration of the Truman Doctrine, and the Marshall Plan. It saw the invention of the transistor and Chuck Yeager's breaking of the sound barrier over the dry lake at Muroc.

One of the first tasks of the new Air Force was to explain reports of mysterious craft—possibly craft from distant stars.

They were as shiny as mirrors, pilot Kenneth Arnold reported. He saw nine objects near Mount Rainier, Washington, on the afternoon of June 24, 1947, in loose formation, shaped like boomerangs or flying wedges, moving at tremendous speed.

After landing in Pendleton, Oregon, Arnold described his sighting to Nolan Skiff, a columnist for the *East Oregonian*, and told him how the objects "flew like a saucer would if you skipped it across the water." The Associated Press picked up Skiff's story, and in its version the objects changed from flying like a saucer into "saucer-like" objects, then into "flying saucers."

The flying saucer would come to inhabit many of the the dreams of the postwar era, focusing fears and hopes like the lens whose shape it shares, reflecting the wider culture like its mirrored surface. Nothing says more about its origins than the birth of its name in the press. For the image of the saucer was about to become a new kind of mythological figure, a Hermes or Puck, a unicorn or leprechaun, that flourished not in oral tradition but in the mass media. The first folk emblem to emerge from the realm of technology, it turned into the most flexible sort of cultural icon, with overtones ranging from the cosmic—dark visions of potential invasions—to the comic—a thousand magazine cartoons with stubby saucers piloted by little green men.

In the days after Arnold's sighting, dozens of additional reports flowed in from around the world. In July, the Air Force boldly issued a press release claiming the "capture" of a flying disc, at Roswell, New Mexico, then decided that the object had in fact been a weather balloon. The Roswell story quickly dropped from the headlines—to be reexamined only decades later—but within two months, polls showed that 90 percent of Americans had heard of flying saucers.

Arnold at first thought he had seen advanced military aircraft. The flying saucer was "discovered" amid almost daily announcements of wildly new technologies and rising tensions, which in the new atomic age threatened the end of the planet. The saucer became a fact of life, like the nuclear threat, and soon it was common enough to be treated lightly. Billy Ray Riley and his Little Green Men had a hit record with the rockabilly number "Flying Saucer Rock and Roll," and by 1957 there was a new toy in American backyards: the saucer-shaped Frisbee, product of the Wham-O company.

————

The embodiment of airpower in its new guise as the atomic deterrent force would be the Strategic Air Command, and its leader Curtis LeMay.

LeMay took charge in October 1948 and declared the SAC a shambles, with untrained crews who couldn't hit their targets. He staged a mock bombing attack on Dayton, Ohio. It was a dismal failure—most crews missed. LeMay called it the darkest day in the history of airpower. He proceeded to get the SAC into shape.

He gave SAC its motto: "Peace is our profession." It said so on its seal, a shield bearing an armored hand glinting like an airplane against a blue sky—an image like a knight painted by Piero della Francesca. But the SAC seal had three lightning bolts and only one olive branch. LeMay's premise was: We are at war already. Since the next war would be one of deterrence, won or lost before it started, we were in effect already fighting World War III. So LeMay kept some of his planes in the air at all times. All were designed to scramble quickly, with a red button for one-touch start-up inside the nose wheel wells where you boarded the plane, for a kind of Le Mans start.

He had no hesitancy about striking first if attack seemed imminent. With every passing year, the margin of advantage for the United States grew smaller. SAC's advantage, LeMay said, was a "wasting asset." It seemed crazy to him to let the other guys strike first. "Hit 'em with their pants down," as George C. Scott urges, portraying the general in Dr. Strangelove modeled after LeMay.

In June 1950, SAC staged an exercise involving dozens of bombers that targeted Eglin Air Force Base. In Mission with LeMay, the autobiography LeMay wrote with MacKinlay Kantor,[3] he described his methods of constant practice: "We attacked every good-sized city in the United States. People were down there in their beds, and they didn't know what was going on upstairs. By the time I left SAC, . . . every city in the United States of twenty-five thousand population or more had been bombed on innumerable occasions. San Francisco had been bombed over six hundred times in a month."

———

LeMay had an obsession with security and a fear of sabotage. He gained national publicity when he staged a surprise visit to a SAC hangar and found the security guy eating lunch. "I saw a man guarding our planes with a ham sandwich," he said. He had crack Air Police patrolling SAC bases, like the commando units depicted in Dr. Strangelove. He dispatched trained "penetrators" to plant notes that said, "This is a bomb." This obsession shows up in the film Strategic Air Command, in which mild-mannered Jimmy Stewart goes back to the Air Force and is baffled by the rough security checks at the base gate. It's "Mr. Smith Goes to Omaha," and it may be one of the least convincing military movies ever made.

Sometimes they would paste a baby picture or animal picture on the ID badges just to test security guards. Once a SAC general found soldiers entering his office to repair phone lines. It took the officer several minutes to remember that the Air Force used outside repair people. He drew his automatic before the intruders had time to deposit the slip of paper that read, "This is a bomb."

The Office of Special Investigations penetrators became a regular nuisance to SAC crews. LeMay even had his wife tested by a bogus repairman who tried to penetrate the general's residence.

——

SAC's headquarters was at Offut Air Base in Omaha, Nebraska, formerly a dreary Army post. The location had been chosen carefully: By the Great Circle route, it was as far from the bases of Soviet bombers as possible. Like railroad towns, Offut and the other distant SAC bases at Rapid City or Minot quickly turned into American dream towns. LeMay made SAC a housing developer, creating whole new communities around the bases, green-grass Levittowns under blue skies. He set up hot-rod shops on SAC bases to improve morale. The cars raced on the runways. It was Pax Atomica, as LeMay liked to call it.

——

"Do you realize how many babies are born in SAC each month?" said Jimmy Stewart, as a B-36 pilot in *Strategic Air Command*. I had been one of those babies. I grew up on a SAC base.

I grew up with the religion of airpower. I must have been but three or four when my mother brought home a model of the B-29 on which my father had flown, all silver, with burgundy prop and tail tips, and I learned that the airpower that had won the last war was there to prevent the next. Like many of the Interceptors, I had "imprinted" on these aircraft as a child, the way Konrad Lorenz described the imprinting nature of goslings. The B-36s overhead were just a larger, clunkier version of the B-29; the B-52s and B-47s and B-58s would continue the evolution.

My father figured as a heroic warrior of airpower. Family myth segued neatly into the national myth that arrived on our primitive black-and-white TV set via Walter Cronkite and the program *The Twentieth Century*: how eager American youths from small towns across the country were sent for training to the new bases set up far from the vulnerable coasts.

Then Air Power had shuffled the trainees into ethnically mixed all-American crews—the kid from Brooklyn, the guy from Texas, the farm boy and the city boy—that would fly from Wichita to Khartoum and Bombay, to China and Guam, and eventually over the Imperial Palace in Tokyo. My father had bombed Tokyo in LeMay's great firestorm, then been shot up over Osaka, left blind, with his right arm crooked and bent.

His left compensated; from my earliest days I thought it looked like the arm on the baking soda box. Decades later, bits of shrapnel were still working their way out of his skin.

The B-36, the flagship of SAC during the 1950s, was something of a turkey, slower and with less range than promised. Originally designed in 1941 to reach Germany from the United States in case England fell, it first flew in 1946. It was jokingly called "aluminum overcast" for its huge size. A mechanically ragged airplane, it was saved by its abundance of engines. There was another joke about it: "Pilot: Feather four. Engineer: Which four?" The bomber's big, slow propellers emitted a distinctive whump-whump sound. One pilot recalls that it sounded like a streetcar rumbling toward takeoff.

It was huge, with six pusher-prop turbojets set along its wings so thick crewmen could scramble out to work on the power plants in flight. But the dome and the bulbous nose gave the plane a stupid, brontosaurian look. In flight, the great glass-domed turtleback canopy atop the bomber was often filled with blue smoke from the cigars the pilots felt free to smoke on long flights because LeMay was rarely without his own stogie. Sometimes a cigar is just a cigar, but in SAC it was a symbol of jaunty esprit, an accent of élan on the way to the end of the world.

Trophies given to the winning crew in a SAC competition one year were ashtrays with a B-36 mounted on their rim, circling the smoking ashes beneath.

―――――

SAC was staffed by callow youth and bomber vets, "the Blue Sky Boys," who had pounded Germany and Japan with Flying Fortresses and Superfortresses and who got the nod in 1948 to deliver the big ones. SAC's job was to routinize Doomsday, to bureaucratize Armageddon. They stayed airborne twenty-four hours a day.

SAC's Cold War was a new kind of war, but LeMay still needed targets. He needed them to etch into three-dimensional Lucite templates for the radar bombsights of his bombers. He needed them to flesh out his Strategic Library Bombing Index. He needed them to shape the SIOP, the sinister acronym for single-integrated operating plan—the blueprint for nuclear war.

LeMay needed targets because he alone controlled them. Neither the joint chiefs nor the president knew the targets in case of nuclear war.

LeMay kept the information to himself until the early sixties. And since there were no locks, no presidential codes for the weapons, his bombers could have launched a nuclear war on his authority alone.

LeMay feared dilly-dallying politicians: He wanted to "hit 'em with everything we've got" at the first signs of any massing of the bombers he was sure the Soviets were rapidly building. But he had very little information. The Soviet Union was a great black empty space. SAC was still using German maps of the country from World War II. Human agents had little success. They might manage to pass for ordinary Soviet citizens, but ordinary Soviet citizens had virtually no access to the areas and targets desired. Reconnaissance versions of the B-29 had skirted the perimeter of the Soviet Union since the end of World War II. A variety of electronic listening and air-sample programs had been in continuous operation.

Other ideas floated around. In the early fifties a forward-looking officer at Wright-Pat had taken a look at new engines and wings and realized it might be possible to fly above radar. Maj. John Seaberg began Project Bald Eagle, developed to create a high-flying spy plane. Specs were issued, proposals advanced, but nothing came of it.

Several balloon programs had been used to spy; one was Mogul, the secret program later officially asserted to have been the source of the Roswell "saucer" wreckage, aimed at sampling potential fallout from Soviet atomic weapons.

The most ambitious balloon program carried cameras: Project Genetrix, aka Weapons System 119L, launched polyethylene balloons high into the jet streams. It operated under the cover story of weather research and the code name Moby Dick. It involved five launch sites and ten locations for tracking, and the Soviets protested as soon as the first flight was made, in January 1956. Almost five hundred balloons were launched; some were shot down, many were lost, and only forty produced any useful photos. The program ended with the humiliating spectacle of captured balloons displayed in Moscow's Gorky Park as evidence of imperialist treachery.

LeMay also enlisted the help of the British for a less confrontational approach and supplied them with planes, Canberra bombers adapted for reconnaissance. They fared poorly. The historian Richard Rhodes records that one pilot from those missions, looking out of his cockpit, realized what a difficult task it would be to find anything in the vast landmass. It looked, he said, like "one large black hole." Some of the Canberras returned full of bullet holes.

When President Eisenhower, in his Open Skies proposal, suggested that the United States and the Soviet Union should allow each other free reconnaissance overflights, the Soviets were suspicious. They rejected the proposal immediately.

At the height of Cold War tensions, in 1956, LeMay sent a fleet of RB-47s over Vladivostok at noon without approval from his commanders. They took pictures boldly, brushing off the few MiGs that rose to intercept them. He could easily have started a war with such a flight. In fact, there is much evidence he regretted not doing so.

In the spring of 1953, a top-secret RAND corporation study pointed out the vulnerability of SAC bases to a surprise attack by Soviet long-range bombers. That August, just nine months after the first American blast, the Soviets tested their first hydrogen bomb. LeMay thought they would be ready to attack by 1954—the year of "maximum danger." Others, however, believed that salvation from impending nuclear holocaust could come only from the intervention of agencies from beyond the threatened planet itself.

8. "Something Is Seen"

Driving from Los Angeles to Las Vegas one day, I took the southeastern route and made a diversion to catch a glimpse of the shrine of the saucers. Near Twenty-nine Palms, where the Marine Corps had its vast desert training ground, stands a white, domed building called the Integratron. It reminded me of the dome the Air Force had built atop Bald Mountain to provide a commanding view of Dreamland.

Here, in April 1954, five thousand people attended "The World's First Interplanetary Spacecraft Convention," and all the important figures from the flying saucer world were there, including contactees George Adamski, Daniel Fry, Truman Bethurum, and Orfeo Angelucci.

The event's organizer was George Van Tassel, a former aircraft mechanic at Lockheed. In 1947, Van Tassel had leased the airstrip at Giant Rock, named for a huge boulder in the desert east of Los Angeles. A German spy was rumored to have hidden beneath the boulder during World War II, and Van Tassel dug rooms there and established a café restaurant for fliers and the less frequent auto tourists. He furnished one chamber with sofas and couches and even a piano. He found it just the place to make contact via telepathy of the "omnibeam" with the space people and "etherians," and he established a Council of Twelve that provided "the first mental contact" from Ashtar, commandant of a space station. Under direction from his voices, Van Tassel began building the Integratron, a dome-shaped structure that would focus spiritual forces with which to prolong life and make possible both antigravity transportation and time travel. Left unfinished at Van Tassel's death, the building is now derelict, its paint peeling. After 1955, attendance at the annual "Saucerian conventions" would slowly decline.

One of the most charismatic figures at the first Saucerian convention was another Lockheed alumnus, Orfeo Angelucci. One attendee noted that he was the only one of the contactees she could really imagine on a spaceship, describing him as "a small, slender, almost fragile man" with "dark, wavy hair, trusting eyes, and a delicate, semi-ascetic face . . . frequently reminiscent of a saint's head by da Vinci . . . The softness of his voice reflects the quality of quiet perseverance."

In 1952 Angelucci was hired at the Lockheed factory in Burbank as a mechanic. He was a nervous man, often in ill health, who suffered from what he believed was "constitutional inadequacy." He felt small. He had been born in Italy, was not very well educated, and was not at all sure he was going to make it in the bustling get-ahead southern California culture of the early fifties.

He had pretensions: He believed his wife was a distant relation of the storied Medici family. He fancied himself something of a thinker. He had seen a UFO in 1946, and pondered its meaning, and for several years had been working on an ambitious philosophical treatise he called "The Nature of Infinite Entities." Its subject matter was "Atomic Evolution, Suspension, and Involution, Origin of the Cosmic Rays."

Coming off the late shift at Lockheed on the night of May 23, 1952, Angelucci felt unwell; his skin prickled. Driving home, he saw a strange red light hovering above the highway and "pulsating." It then shot off to the west at about a 30- or 40-degree angle into the sky and vanished. In its place, two smaller green orbs like green fire hung in front of him.

He pulled off to the side of the road. He understood that the green orbs were some sort of transmitting and receiving devices. They moved closer and formed a kind of 3-D film screen between them; two faces, a man's and a woman's, appeared on the screen, and Angelucci heard a voice, declaring them "friends from another world," "etherian" beings who had come to Earth.

They asked him if he remembered seeing a UFO on August 4, 1946. He replied that he did. He was suddenly very thirsty; a crystal cup appeared on the fender of his car and he drank from it. The beverage tasted wonderful.

The voice told him that distant civilizations were concerned with man's "spiritual progress," which had not kept pace with its material progress. "Weep, Orfeo," they said. "For all its apparent beauty, Earth is a purgatorial world among the planets evolving intelligent life. Hate, self-ishness, and cruelty rise from many parts of it like a dark mist."

People on Earth, he was told, did not appreciate each other. But the etherians did. "Every man, woman, and child is recorded in vital statistics by means of our recording crystal disks. Each of you is infinitely more important to us than to your fellow earthlings." They warned of a great cataclysm that would strike Earth in 1986 if changes were not made.

Angelucci would have more encounters—at the Greyhound bus terminal and in the dry bed of the Los Angeles River. In 1955, he published an account of his adventures with the people from space in a book titled *The Secret of the Saucers.*

As he recounted it, on July 23, 1953, he again felt unwell and stayed home from work. In the evening he took a walk, and on his way home, in a lonely place, he felt a "dulling of consciousness." There, in the bed of the Los Angeles River, he saw an "igloo-shaped" spaceship, like a "huge, misty soap bubble." A door appeared in the bubble craft. He entered and found himself in a vault about eighteen feet high, lined with some "ethereal mother-of-pearl stuff." He saw a chair of the same mother-of-pearl, and when he sat in it, it seemed to mold itself to his body.

A humming noise puts him into a semi-dream state. He is carried off into space, and sees the earth from a thousand miles away. He passes a UFO a thousand feet long, made of a crystalline substance and emitting music and images of harmoniously evolving planets and galaxies. The UFO is equipped with "vortices of flame" that serve as both propellers and some mode of telepathic contact. He wakes to find a mark on his chest about the size of a quarter: a circle with a dot in the center that he decided was a symbol "of the hydrogen atom."

In September 1953, he would spend a week in a semiconscious "somnambulistic state." He awoke and recalled, as if from a dream, that he had been spiritually transported to another planetoid and met Orion and his spacewoman friend Lyra. He learned that he had himself been a spaceman in an earlier life, named Neptune.

———

One avid reader of Angelucci's book was Carl Jung, who had been paying attention to flying saucers since 1949. He saw them as an example of a modern myth being born before his very eyes, and nowhere was the process clearer than in the accounts of contactees, people like Angelucci who claimed not only to have seen flying saucers but to have spoken with their crews and even flown on them.

Angelucci's dreamy account fascinated Jung. Perhaps it was the naïve, almost old-fashioned quality of the experience, as shown even in the design of the saucer's crystal walls and mother-of-pearl interior, or the mythological and archetypal overtones of the author's name. *Orfeo:* Orpheus, the poet. *Angelucci:* angel of light.

To Jung, Angelucci's story offered a clear example of the process of a UFO sighting emerging from a disturbed spirit. He saw Angelucci's visions, as he understood all saucer sightings, as the expression of a wider cultural unease and disturbance. Orfeo was reaching a solution to a problem by something like the workings of a dream.

"As our time is characterized by fragmentation, confusion, and perplexity," Jung declared, "this fact is also expressed in the psychology of the individual, appearing in spontaneous fantasy images, dreams, and the products of active imagination."

In 1958 Jung published his own quite strange book about UFOs called *Ein moderner Mythus: Von Dingen, die am Himmel gesehen werden*, or, as translated literally, *A Modern Myth: Of Things That Are Seen in the Sky*. The English publishers offered instead a more marketable title: *Flying Saucers: A Modern Myth of Things Seen in the Sky*.

In the introduction Jung refers to himself as an "alienist"—the nineteenth-century term for a doctor who treats the insane, from the French *alieniste*. For Jung, who analyzed UFOs in their relationship to fantasy and interprets a number of UFO dreams, the saucers spring from the cultural state of affairs of the fifties—of the bomb, the Cold War, McCarthyism, and the resultant fear and confusion. He attributed much of the UFO phenomenon to a wide sense of unease in the culture, and there was, he believed, a tendency for underlying emotions to "manifest" themselves in observations of real or imagined things. "Universal spiritual distress" causes us to see archetypal circles in the sky.

Jung did not attempt to decide whether flying saucers were real or not. He treated them as "symbolical rumors." For him the saucer was an archetype in the making, an icon "weightless as thought." He wrote, "The round shape, the saucer, is the shape of the center, located deep in the collective unconscious. Such an object provokes, like nothing else, conscious and unconscious fantasies."

There was no question for Jung that something had been seen, that observers had *seen something*, but whether reality or illusion he was not sure—nor did he think it very important to distinguish them as such.

"One often did not know and could not discover where a primary perception was followed by a phantasm," he wrote, "or whether, conversely, a primary fantasy originating in the unconscious invaded the conscious mind with illusion and visions." In essence, Jung was saying, they might be real and they might not. But he saw the archetypes living in a kind of unconscious symbol language we all possessed, and he turned to ancient mythology, religious tracts, astrology, astronomy, and alchemy for his primary comparisons.

Jung had not discussed the fact that the lore had moved beyond the old oral and written sources, beyond the campfire, the village square, the learned tome or tract—and to the modern news media. Increasingly, the mass media had become the medium where his beloved archetypes now lived and mutated, like organisms in a lab vial.

Jung was baffled when his first statement on UFOs was picked up by the popular press as a sign that he *believed in* flying saucers. He released a clarifying statement to U.P.I., and was surprised when it was given far less distribution than the earlier statements.

Had Jung looked more closely at the history of the flying saucer sightings in his book, he might have noted how vital the role of the press had been from the very beginning. Kenneth Arnold's 1947 ur-sighting would never have set off the saucer obsession had not reporter Nolan Skiff seized on the image of the skipped saucer; from it was coined the catchy *flying saucer*—a phrase Arnold had never uttered—sent out on wire services all over the world.

Jung concluded only that "something is seen but it isn't known what," admitting that this "leaves the question of seeing open."

He was more interested in what caused the "seeing." The round shape, such as that of the saucer, is located deep in the collective unconscious, he declared. It is the mandala, the rotundum, age-old, deep, and powerful as the lenticular shapes of galaxies. God is often described as round, a circle with no edge and no center, or as a watching eyeball. "The center is frequently symbolized by an eye," as in the all-seeing eye of the conscience.

When the common center cannot hold, the round shape appears as a wish, a response to "the fears created by an apparently insoluble poetical situation which might at any moment lead to a universal catastrophe," Jung wrote. "At such times men's eyes turn to heaven for help, and marvelous signs appear from on high." A rationalistic world grounded in science and technology, a world of "statistical or average truths," perhaps unable to deal with these things, creates "an insatiable hunger for anything extraordinary."

And if there really were aliens here? Then, Jung states, we would be in the position of a primitive tribe dominated by white Western power. The reins of power would be wrenched from our hands. "As an old witch doctor once told me with tears in his eyes, we would have 'no dreams anymore.' The lofty flights of our spirit would have been checked and crippled forever, and the first thing to be consigned to the rubbish heap would be our science and technology." In such a situation, we would just roll up the Iron Curtain and get rid of our weapons. Jung prefigured Ronald Reagan's oft-cited declaration to Mikhail Gorbachev that should alien invaders appear, our two countries would learn to get along soon enough.

Of course the question of the reality of the saucers remained. Jung left a big hole of possibility, a portal for the New Agers who would grab onto his ideas years later. The notion of one thing causing another was a narrow, rationalistic view, Jung argued, rejecting it in favor of the explanation that things happen in synchronistic "acausal, meaningful coincidence."

Ultimately, Jung insisted on interpreting the world as a set of symbols, not of realities, of seeing rather than knowing, of "symbolical rumors," of lore—dreams of the collective unconscious. But where did the collective unconscious reside? In the absorbed zeitgeist of strange characters like Orfeo? In the newspapers, the tabloids, the magazines; on the wire services; in the movies? Hadn't *The Day the Earth Stood Still*, a popular and now classic film from 1951, brought essentially the same message as Orfeo's "etherian" visitors? And could not print or film have also brought that message directly to Orfeo, a couple of years before he saw the orbs pulsating?

———

Besides Orfeo Angelucci, the best-known contactee at the first Saucerian convention was George Adamski. He had fought with the cavalry down on the border during the Pancho Villa unpleasantness. In the early thirties, he established a Tibetan temple in Laguna Beach, one of whose virtues was that its status as a religious institution meant dispensation from the rigors of Prohibition. If repeal had never come, Adamski would later say in an unguarded moment, he might never have gotten into "this saucer crap." He moved to the slopes of Mount Palomar and began trying to photograph the saucers. In his 1955 book, *Inside the Space Ships*, he told of being taken on board flying saucers by aliens with mythological names, and he reported that he spoke frequently to his "Space Brothers."

Truman Bethurum, author of the 1954 book *Aboard the Flying Saucers*, reported that while laying asphalt in the desert in July 1952 he saw eight or ten small spacemen. They took him aboard their spaceship, where he met its captain, Aura Rhanes, a female he described as "tops in beauty," from the planet Clarion. Again, the burden of the message was a warning against nuclear weapons and of the need for love.

Daniel Fry's 1954 book, *The White Sands Incident*, prefigures elements of the Roswell and Area 51 stories but with a wholly different tone. Fry worked for Aerojet General at the White Sands rocket test site. On a remote corner of the base, on July 4, 1950, he said, he saw a flying saucer land. From inside, a voice belonging to a visitor called A-lan invited him for a ride to New York and back. In 1955, Fry published *A-lan's Message to Men of Earth*, this time based not on a direct encounter but on "a voice inside my head." Like many of the contactees, he veered toward mysticism, and he tied the saucer tales to classic prewar obsessions with the ancient continents of Lemuria and Atlantis. After a great conflict between the two, Fry suggested, the survivors had fled to Mars.

————

There was a pattern to the lives of these contactees: Almost all had come from the Los Angeles area—and had worked on the edge of the aeronautic industry. Their accounts share a tone and a language. They have been taken aboard saucers, not with the menacing experimental intent described by later abductees, but in a naïve, friendly way. The aliens are friends, "Space Brothers," who address the contactee as "pal." Unlike the abductees who would dominate the youfer lore of the 1980s and 1990s, the mood is not one of manipulation but of wonder, even enlightenment. The ruling spirit is Klaatu, the alien in *The Day the Earth Stood Still*, who has come to warn us of our own folly, specifically nuclear folly. In some of the accounts there is an old-fashioned, almost nineteenth-century feel, as in Van Tassel's assertion that human beings were the result of beautiful Venusians mating with ugly Earth apes.

Fashions in ufology apparently offer a shadow version of the wider culture. For some the aliens are saviors, for others, invaders.

The first mystery "airships" in the 1890s arrived when the fascination with flying machines and balloons was at its height, a time of urbanization, immigration, and economic depression. While the first "foo fighters" of the 1940s, lights spotted by fighter pilots, were discounted as mere oddities, like the false bogeys on crude early radar, the ghost rockets of the immediate postwar years suggested a fear of attack from the Soviet Union.

The flying saucer craze of the late forties and early fifties—culminating perhaps in June and July of 1952, when Washington, D.C., was "buzzed" by multiple saucers, recorded by ground observers, radar watchers, and airline pilots—marched along in neat parallel to McCarthyism and the Red Scare. (To the Japanese, sociologists argued, Godzilla stood for the assault of the B-29s, their incendiary and atomic bombs.) During the hottest period of the Cold War, the aliens brought contactees a message of peace. But already a darker theme of cover-up was emerging, in the charges of leading UFO propagandist Donald Keyhoe that "silencers" were at work and the government was keeping the truth a secret.

Race sometimes emerged as a theme of UFO stories in the sixties, and the theme of government cover-up—a shadow of the assassinations, Vietnam, and the Pentagon Papers—grew stronger in that decade. The national humiliations of Watergate and Iran coincided with the cattle mutilation stories of the seventies.

Close Encounters of the Third Kind featured François Truffaut playing a thinking man's UFO expert, based on the UFO researcher Jacques Vallee, who echoed Jung's arguments about considering sightings on their own terms and skirted the issue of real existence. But in the end of the movie, real saucers *do* appear.

Fashions in ufology changed in the eighties, when E.T. (1982) was understood as a fable for childhood. Children, like aliens, are new to the planet, with innocent assumptions and virtually no knowledge about how life is lived here on Earth.

The eighties craze for abduction stories was in keeping with the cultural trends of the rest of the decade. Its sexual and personal obsessions—I was taken because I was special, I was abused—tied in with talk-show psychology, itself an emblem of the times.

In the eighties, too, Stealth created its own shadow culture in the Bob Lazar story. The F-117 looked like a flying saucer when viewed head on—and for sound technical reasons. Ben Rich would write of the design of the fighter, "Several of our aerodynamics experts, including Dick Cantrell, seriously thought that maybe we would do better trying to build an actual flying saucer. The shape itself was the ultimate in low observability. The problem was finding a way to make a saucer fly. Unlike our plane, it would have to be rotated and spun." This statement was widely cited by both those merely curious about flying saucers and those firmly convinced of their existence.

The secrecy around Stealth helped nurture rumors that it had been created with the assistance of alien technology; one saucer organization noted that when a still secret Stealth fighter crashed in the summer of 1986, the whole area was cordoned off and cleaned up just as the Roswell crash and other "recoveries" had been.

————

The eras of changing fascinations in the UFO culture suggest periods in fashion or movements in art. And many of the contactee visions reminded me of what is called outsider or visionary art. In the paintings of these socially marginal and untrained artists—"kooks" or "loons," in the later parlance of the Interceptors—flying saucers appeared as frequently and naturally as angels or Jesus, or 727s and locomotives. These artists often actually paint UFOs. Like many of the contactees, they not only see visions but hear voices, inspiring them to paint landscapes from other planets or construct saucer shrines, even landing pads.

Many of these images possess a dreamy, otherworldly quality, like Angelucci's prose, in which Tiny's Cafe in Twenty-nine Palms turns into a magic chamber where he sips amber. Others share the intrigue in detailed alternative engineering and dissident cosmology with the saucer buffs, who look to Nikola Tesla and Townsend Brown as alternate-world heroes of the technology of conspiracy.

————

Van Tassel's Giant Rock "spaceport," it turns out, was merely one of many smaller offshoots. I came across a book that documented a world of such people who built UFO detectors and landing sites for saucers. These were believed to be the vehicles of angels or aliens, or both. Douglas Curran, the book's author and photographer, recounts that the title, In *Advance of the Landing: Folk Concepts of Outer Space,* had come to him in a dream. Curran, like Jung, found that when he tried to approach the saucer sighters and cultists as a folklorist, there were those who still pulled him aside and earnestly asked, But do you believe? This suggested just how close such folk cultures were to religious sects, which helps explain the shrinelike nature of the places Curran photographed.

Sightings and imaginings, theories and conspiracies—the cultures of Dreamland made up a folklore of its own. Did it matter whether the Aurora airplane or the "alien replicated aircraft" actually existed, any more than whether Hermes actually had wings on his feet? Folklore and superstition begin where science and knowledge end. And knowing stopped at the perimeter around Dreamland.

———

After reading Jung, I became more aware of patterns in the tales surrounding Dreamland. Like UFOs, the actual existence of the flying black triangles (or bats or rays or pumpkin seeds) was a matter of serious debate. The black-plane stories shared a consistency of account and rough detail that made up a corpus of experience. I came to think of it collectively as the Lore.

Today's folklore, or the nearest thing we have to it, is bounded by technical expertise and collective fascination. It lives in a group's language, assumptions, and perspective, in its prides and prejudices. Technical subcultures—sharers of belief in a technology—are paralleled by those with a faith in conspiracy, a hidden order. Could it not be that in an age of technological explanation it took the unexplained to link us together? That in the age of information, it took mystery? Shared professions and shared fascinations had replaced the shared geography of village or town. Sometimes the cultures of technology could seem like cults, and the mechanics of conspiracy theory could seem as complex as science or engineering.

Both saucers and mystery planes had about them the same compulsive gathering of bits of information, the careful construction of databases of sightings, dimensions, aircraft specifications, and numbers. In this regard, both groups resembled the historian Richard Hofstadter's descriptions of conspiracy groups who from time immemorial have built elaborate factual structures from which to launch speculations.

John Pike of the Federation of American Scientists saw the same sort of dynamic Carl Jung had observed among flying saucer buffs at work in the sightings of black aircraft. "Considered as a sociological and episte-mological phenomenon, the parallels between reports of flying saucers and reports of mystery aircraft are striking," he wrote. If, as Jung be-lieved, flying saucers "were a response to the deep cultural anxieties of a society threatened with sudden nuclear annihilation," then couldn't mystery aircraft be a response to economic challenges and the decline in fortunes of the aerospace industry, whose future the end of the Cold War had made uncertain?

"Belief in the existence of marvelously capable and highly secret air-craft resonates with some of the deeper anxieties of contemporary American society," Pike went on. "Aviation has long been one of the dis-tinguishing attributes of American greatness, from Kitty Hawk to Desert Storm.

"It would be reassuring to believe that concealed in the most hidden recesses of the American technostructure were devices of such miracu-lous capabilities that they will astound the world when at last they are re-vealed and will restore America to its rightful station of leadership."

The saucers might save us from the Cold War; the black aircraft could save us from its aftermath.

9. Ike's Toothache

Not long before the Saucerian convention at Giant Rock, the president of the United States came to nearby Palm Springs to relax. He had made the eight-hour flight out to California on his Lockheed Constellation, named the *Columbine* after a wildflower he loved from his prairie childhood. On Saturday, February 20, 1954, Dwight D. Eisenhower was enjoying a golfing vacation at Smoke Tree Ranch as the guest of golf partners Paul G. Hoffman, chairman of the Studebaker Corporation, George Allen, an insurance CEO, and Paul Helms, president of the Helms Baking Company. He rose early, met the press at eight-thirty to announce he had signed twenty-three bills, and made comments supporting his nominee for Chief Justice of the Supreme Court, Earl Warren. He spent the day playing golf. But that evening, after dinner, Ike disappeared.

One can easily imagine the press corps, happy for some time out of the Washington winter, sitting around their Saturday night card game in the nearby Mirador Hotel and getting irked when Eisenhower did not return from dinner as scheduled. Could he have had a heart attack? Was a world crisis brewing? What one correspondent would call "journalistic mob hysteria" seized the press when Merriman Smith of the U.P. dispatched an alarming report that the president had been taken away for "medical treatment." The rival A.P. took it another step: The president was dead, it declared in a hastily retracted bulletin.

James Haggerty, the press secretary, was called out to make an explanatory statement. The wild rumors were quickly put to rest. "During [the president's] evening meal, the porcelain cap on one of his front teeth chipped off," *The New York Times* reported. "Mr. Helms took him to Dr. F. A. Purcell, a dentist, who replaced the cap. When the president goes to church tomorrow morning, his grin will look the same as ever."

The reporters grumbled about a toothache being turned into an international crisis, but during the hours of Eisenhower's absence, a legend was born: The Lore would record that he was secretly flown to Muroc, soon to be Edwards Air Force Base, to meet with aliens and view recovered flying saucers. Eisenhower's dental mishap, like the crumb of cheese that grows into Scrooge's nightmares, would grow into a whole fabric of conspiracy theories that will eventually end up in Dreamland.

The next morning Eisenhower took his wife and mother-in-law to the Palm Springs Community Church, his repaired grin inspiring crowds to political-rally warmth. In the sanctuary, the minister praised Ike and Mamie's spiritual example, their witness to Christian principles and religious conviction. One aspect of his religiosity was that Ike did not play golf on Sundays.

Had Ike made that trip, met those aliens, could the grin indeed have looked the same as ever? It must have been a moment of profound philosophical reexamination for the former general. According to one account, the aliens "kept disappearing, causing him embarrassment." Did he wonder where to focus his attentive gaze, his welcoming remarks? Could this man have indeed disappeared between dinner and breakfast to view hidden saucers and meet with aliens and then sailed off to listen happily to that sermon?

The idea of a trip to Muroc is hard to buy. The president would have had to fly to leave himself any significant amount of time at the base. He would have lost a lot of sleep.

————

The legend of "Ike's toothache" was established in UFO lore as a result of a letter written in April 1954 by Gerald Light to Meade Layne. No one has much of an idea who Light was, beyond the fact that he was an adherent of a spiritualist organization called the Borderlands Foundation, founded in 1945 by Layne to explore "realms normally beyond the range of basic human perception and physical measurement." Publishing works by Charles Steinmetz and *The Etheric Formative Forces in Cosmos, Earth and Man* by Dr. Guenther Wachsmuth, Borderlands was dedicated to investigations of "ether ships," Vril energy, radionics, and dowsing. It stood somewhere between the Theosophist groups then influential in Los Angeles and today's New Age groups. Layne himself had written on the saucers, which he called "ether ships" or "aeroforms," tying them to the Kabala and other mystical writings.

Light's letter has become a classic of the Lore, a record of suspicion emerging from enthusiasm, excitement mingling with dread.

> My Dear Friend—
> I have just returned from Muroc. The report is true—devastatingly true!
> I made the journey in company with Franklin Allen of the Hearst papers and Edwin Nourse of Brookings Institute and Bishop MacIntyre of LA (confidential names, for the present, please).
> When we were allowed to enter the restricted section (after about six hours in which we were checked on every possible item, event, incident, and aspect of our personal and public lives), I had the distinct feeling that the world had come to an end with fantastic realism. For I have never seen so many human beings in a state of complete collapse and confusion as they realized that their own world had indeed ended with such finality as to beggar description. The reality of "otherplane" aeroforms is now and forever removed from the realms of speculation and made a rather painful part of the consciousness of every responsible scientific and political group.

During his two-day visit, Light went on, he saw five different types of aircraft "with the assistance and permission of the Etherians."

The notion that he would have been included with such well-known figures as the Hearst columnist and the bishop is self-congratulatory, and the tone is a strange combination of sermonly seriousness and offhand weirdness:

> President Eisenhower, as you may already know, was spirited over to Muroc one night during his visit to Palm Springs recently.
> Mental and emotional pandemonium is now shattering the consciousness of hundreds of our scientific "authorities."

"Pity" was what he felt watching "the pathetic bewilderment of rather brilliant brains struggling to make some sort of rational explanation." For himself, he said, he had long ago entered "the metaphysical woods."

> I had forgotten how commonplace such things as the dematerialization of "solid" objects had become to my own mind. The coming and going of an etheric, or spirit, body has been so familiar to me these many years I had just forgotten that such a manifestation could snap the mental balance of a man not so conditioned.

Light's letter reads like the most clever sort of propagandist document—one whose real message is oblique. While designed to be read by someone outside, it speaks as an insider: Light would not have to define "etheric" for his pal Meade Layne (a name smarmy enough for a character from a Chandler novel). He drops the names of his companions (an unlikely bunch) and describes a thoroughgoing background check that only someone unfamiliar with the military could imagine. Such signs mark his letter as an effort to shift the discussion of flying saucers into the territory of the Borderlands and other spiritualist groups. The flying objects were not from Mars or Zeta Reticuli but from a "higher plane," "a different dimension."

The leaps of speculation implicit in references to Eisenhower's "secret trip" slip in almost unnoticed. Thus uncertainty or secrecy mutates into fantasy: If the president catches cold, the stock market may get pneumonia; if the president has a heart attack, the whole Cold War balance trembles. When the president got a chipped tooth, in this year of maximum danger, consternation ensued. From a tiny chip, a crevasse of speculation could grow.

———

But it was too late for the conspiracists to be disarmed. The first few months after their advent in 1947 was probably the last time that an air of open-mindedness about flying saucers was sustained. The lines of opposition had not yet hardened between private researchers and government. The Air Force, just established as an independent service, had not grown disgusted with the question. Fear had not yet overwhelmed curiosity. A variety of ideas were in play, and speculation was neither stifled nor rampant. Theories of government cover-up had yet to take root. The question, in short, was still open.

On September 23, Gen. Nathan Twining, commanding general of the Air Materiel Command at Wright Field, wrote a secret memo to Brig. Gen. George Schulgen, chief of the Air Intelligence Requirements Division at the Pentagon, about the flying saucer question.

Twining's memo offered what seems a reasonable and open-minded listing of possible explanations for the UFOs: They are a secret U.S. craft, or a secret Soviet system, perhaps developed with the aid of German scientists (shades of future theories). They are an unexplained meteorological or atmospheric phenomenon, or—and this was not ruled out—craft from another star system. Indeed, he added, "It is the considered opinion of some elements that the object may in fact represent an interplanetary craft of some kind.

"The phenomenon is something real and not visionary or fictitious," Twining concluded, in words that would be cited again and again. There was recommendation for further study and a suggestion, later explicitly rejected for reasons of cost, that interceptor fighters be kept on alert to shoot down UFOs.

In December 1947, the Air Force set up Project Sign to track the saucers and other UFOs and determine their nature. But in 1949 the name was changed to Project Grudge, an unconscious symbol that the attitude of the Air Force had quickly turned to irritation. It hated dealing with the UFO problem, the press, the watchers, the nuts. It was uncomfortable with the notion that objects might be able to fly so easily through its air defenses. It felt disarmed dealing in areas where hard evidence was hard to get. Most of all, it wanted to be rid of the problem. The Air Force, a joke had it, wished the saucers swam instead of flew so that they would become the Navy's problem.

In 1948, a report of Project Sign, called "Estimate of the Situation," was completed. Neither chief of staff Gen. Hoyt Vandenberg nor Twining found it acceptable. Deciding that the evidence did not support the conclusions, Vandenberg ordered it destroyed, and the "Estimate" was not made public. Years later, when UFO researchers asked for a copy, neither the military nor the civilian agencies could or were willing to provide one. This lapse would be cited in support of the argument that there was a cover-up. Quotations ostensibly from the never-issued document made reference to descriptions of material that sounded a lot like the Roswell wreckage, which believers saw as proof that the Roswell recovery had been part of a cover-up, that the stuff was indeed pieces of a saucer, and that Air Force units were being asked to be on the alert for similar incidents and objects. The cries of cover-up would soon be a dominant note in the debate over UFOs. In January 1950, *True* magazine published the famous Donald Keyhoe story that charged the government with a cover-up of the truth about the saucers. "Estimate of the Situation" would become legendary and leave a legacy of suspicion.

Keyhoe published a book-length argument, *The Flying Saucers Are Real*, in 1950 and soon began to speak of "silencers," Air Force officers or other government agents intimidating witnesses to keep the truth secret. He followed up with *The Flying Saucer Conspiracy* in 1955.

The strand of the Lore charging cover-up and conspiracy spun off from that of the happy contactees almost immediately. An early and recurrent part of it were the "Men in Black." Here one could clearly see folklore crystallizing from real events.

It began with a plane crash, like a science fiction film I remembered from childhood—was it *Target Earth*? A B-26 crashes, and marks in the dirt indicate the movement of an invisible creature. Suddenly, a dead aviator shudders back to life and begins a zombielike walk—a military man possessed by an alien force.

In July 1947, Fred Lee Crisman and Harold A. Dahl, two men from Tacoma, Washington, who claimed to be harbor patrolmen, reported that they had seen a group of doughnut-shaped UFOs near Maury Island and had gathered scraps of one that had crashed. There were intriguing details: Their radio was jammed and strange spots appeared on photographs they took. And a mysterious man in black drove up in a black Buick and told them to keep quiet.

The Air Force dispatched its top flying saucer investigators, Lt. Frank Brown and Capt. William Davidson from the TID—Technical Intelligence Division. They determined that the whole story was a hoax. But when their B-25 bomber crashed on their return home, suspicions immediately arose that someone was hushing things up. The Tacoma newspaper headlined the story SABOTAGE SUSPECTED.

A book by Gray Barker called *They Knew Too Much About Flying Saucers* was published in 1956 and established the Men in Black legend firmly in the Lore. In 1952, Barker claimed, a man named Albert K. Bender organized a UFO group called the International Flying Saucer Bureau. Within months, Bender was visited by MiBs who told him that the government knew and would soon reveal the truth. They persuaded him to dissolve his organization.

In just a few years, the Men in Black story took on detail and showed all the mutability of a traditional folktale. The Tacoma Buick was upgraded in many versions to a black Cadillac. These Men in Black often dressed too warmly; they walked and talked mechanically; they had vaguely Asian features. They could have been aliens themselves, even robots. To folklorist Peter M. Rojcewicz, they suggested the ominous dark men or evil tricksters found in many folk traditions.

The story took a new twist in 1980 when Lowell Cunningham, hearing of the legend in casual conversation, was inspired to create a humorous comic-book version of the tale. It underwent a further twist when the comic book became the basis for a screenplay and, in 1997, a hit film, *Men in Black*.

Neither Cunningham nor Barry Sonnenfeld, who directed the film, had heard of Crisman and Dahl or Gray Barker or any of the origins of the MiB myth. Knowing it only as an urban legend, they felt free to extemporize on it and the film provided a darkly comic rendition of what had begun as sinister. The film was another play on the alien immigrant/alien life-form pun, and the cinematic Men in Black were urbanized agents belonging to a sort of intergalactic Immigration and Naturalization Service. They provided the great service of keeping us safe and happy in our ignorance of the alien presence.[1]

The camou dudes at Area 51 were in some sense imaginative relatives of the Men in Black. Their danger was overestimated; they took on an almost folkloric quality of menace. Sometimes they inhabited the dreams of watchers, like modern-day Greek Furies. So when Gene Huff talked about his feelings of fear and guilt after visiting the perimeter with Lazar, he talked about "the Dream Police"—police in his mind. The camou dudes, like the mysteries of Area 51 itself, came at the end of a long tradition—a legacy of fear. They were shadows of very old figures of menace, just as the mysteries of Area 51 itself touched almost primal fears of the unknown.

10. Paradise Ranch

For Curtis LeMay, 1954 was the year of "maximum danger," the year he believed the Soviets would have more than enough bombs and bombers to hit us and we wouldn't yet have enough of the silver bombers he thought the United States needed to strike back. For the country at large, it was a year of fear—the depth of Cold War paranoia, the high-water mark of McCarthyism. The greatest fear was of surprise attack, an atomic Pearl Harbor. President Eisenhower feared a surprise attack too, a lack of intelligence like the one that had nearly cost him his reputation at the Battle of the Bulge. If LeMay had seemed eager to start World War III, there were others in 1954 who began creating an airplane that could prevent it.

Out of a sense of near national emergency—a desperate desire to see what was going on inside the Soviet Union—the U-2 spy plane was developed. In July 1954, Ike had created the Killian Committee, chaired by MIT president James Killian, and including leading lights of the scientific community as well as military figures, to decide what to do about the danger of atomic surprise attack. Relying heavily on work by the RAND corporation, its report came in the autumn. One of its key recommendations was the development of some sort of aerial reconnaissance to establish the state of Soviet weaponry.

Major Seaberg's "Bald Eagle" proposal for a high-flying spy plane was brought out of the files; aviation contractors were quietly asked for ideas. Among their proposals was one for a Mach 4 plane launched from the back of a fast B-58 bomber. Another called for a ramjet aircraft to be carried to high altitude by a huge balloon, then released. Kelly Johnson of the Lockheed Skunk Works proposed putting long wings on the fuselage of his F-104 Starfighter—and, he promised, he could do it quickly.

To Edwin Land of Polaroid, a key member of the Killian group, the Skunk Works proposal, called the CL-282, seemed the most practical and potentially the fastest way to put cameras over the Soviet Union. On November 5, 1954, Land wrote a memo to CIA director Allen Dulles called "A Unique Opportunity for Comprehensive Intelligence," pushing the Lockheed idea.

> No proposal or program that we have seen in intelligence planning can so quickly bring so much vital information at so little risk and at so little cost.
>
> We have been forced to imagine what [the Soviet] program is, and it could well be argued that peace is always in danger when one great power is essentially ignorant of the major economic, military, and political activities . . . of another great power . . . We cannot fulfill our responsibility for maintaining the peace if we are left in ignorance of Russian activities.

He made another key point: Such a program was also vital in order to avoid "over-estimation" of the enemy—as dangerous as its opposite.

Land was persuasive and obtained Eisenhower's approval. By December 1954, Kelly Johnson had in his hand a contract to produce twenty of the planes for $22 million—all within nine months. The CIA would foot the bill from discretionary—and very much unaccounted for—funds. And it, not the Air Force, would take charge of the project, code-named Aquatone.

———

Soon a strange figure began to be seen in the shops and offices of the Skunk Works. A tall, stooped man, he inevitably reminded observers of a stork. No one introduced this odd Easterner—anyone at Lockheed could tell that he was from back East, so out of place was he inside the yellow hangars in Burbank—except occasionally he was referred to simply as "Mr. B." He would look even more out of place later on the caliche runway at the base at Groom Lake.

Mr. B. was Richard Bissell, a former Yale economics professor who now headed up Aquatone.

Bissell had grown up in comparative privilege in Hartford, Connecticut, where his family owned Mark Twain's old house. "It was a world unto itself," he would recall, full of odd rooms, secret closets, and private balconies—a happy psychic conditioning, perhaps, for the hidden chambers of the intelligence establishment he was to join. As a child he once tossed his teddy bear off the fantail of the Queen Mary and ordered his nanny to retrieve it. As a teenager, he looked out on the Colosseum and the ruins of the Forum and meditated on the nature of empire.

After Groton and Yale, where he turned down admission to Skull and Bones and became an America Firster, dedicated to keeping the country out of the mounting European conflict, he went on to graduate school. He turned to government just in time to become a key figure in one of the unsung but vital logistics battles of World War II. While U-boats roamed in wolfpacks preying on Allied shipping, and the codebreakers back in England labored to defeat them without tipping their hands, Bissell almost single-handedly ran the Allied merchant shipping program. Using a complex system of file cards, he figured out how to turn ships around fast, and what to fill their limited holds with—bombs or oil or coal—and how to get fruit and tea to London. After the war, he drafted the initial proposal for the Marshall Plan.

In Washington after the war, Bissell's imperial thoughts found congenial territory. Dean Acheson has compared the face-off between the United States and the Soviet Union to that between Rome and Carthage. Bissell quickly became part of the influential Georgetown cocktail party set, gathering over martinis to discuss affairs of state with the Rostows and the Alsops, the Grahams—Kate and Phil of The Washington Post—and the Dulleses—John Foster, Ike's secretary of state, Eleanor, a State Department expert in Asian affairs, and Allen, the head of the CIA.

Bissell returned to academia, then moved to the Ford Foundation. When life there grew dull, he gently dropped a hint to Allen Dulles that he might consider work at the agency. He was one of the new generation of logistics and technology experts, the technocrats, who came out of the war effort—for World War II had been a war of logistics. While Henry Stimson, Republican secretary of state, the epitome of WASP privilege, had shut down the famed Black Chamber in 1929, saying, "Gentlemen do not read each other's mail," the coming of World War II had made it necessary for gentlemen to spy, and to fight dirty. The OSS, a precursor of the CIA, had first brought college men into espionage, and, after its founding in 1947, an Ivy League–educated cadre ran the service. The plots to get rid of Castro that would shock a nation when they were revealed at the Church Hearings in 1973 were like others of the era. The OSS and CIA had come up with wacky schemes before, such as dropping pornographic literature on Berchtesgaden, Hitler's retreat, to drive him mad with lust.

After World War II, the Yale crew coach Skip Waltz was the chief recruiter. A few measured words at the boathouse to a team player, a suggestion in confidence, brought dozens of oarsmen into the CIA. But if many of the Ivy League recruits had been "well rounded," promising men of action, Bissell was unusual—an academic, who came to the agency from a foundation. Bissell liked systems, flow charts, tables. He was the champion of the coming thing in intelligence—technology, photint, elint—and it would sometimes put him in conflict with the traditionalists—the humint people. He would take little interest in the actual content of the pictures the U-2 or SR-71 would take.

Bissell was most interested in the psychology of CIA operations, the manipulation of public opinion, the creation of illusory forces rather than the use of actual weapons.

He would argue in his memoirs that the U-2 served as a psychological weapon as well as a reconnaissance tool. It sowed the crucial idea that the United States could overfly the Soviet Union with impunity. Responding to the humiliation their military felt at having been so brazenly overflown, the Soviets insisted they had shot down the plane at 70,000 feet with a single antiaircraft missile. In fact, they likely fired a huge salvo of the missiles and may have even knocked one of their own fighters out of the sky.

Bissell's first success came with the overthrow in 1954 of Guatemalan dictator Jacobo Arbenz Guzmán. That, along with the replacement of Mosaddeq in Iran with the shah, were the CIA's proudest achievements. Accomplished at the behest not just of the president but of the United Fruit Company, the Guatemala operation set the CIA on the course of exotic "covert," "destabilizing" operations.

Significantly, to Bissell's thinking, the operation turned not on military force but on illusion, such as bogus radio transmissions reporting gathering rebel forces—a twist on the old wartime deception of a few soldiers walking repeatedly back and forth along a wall to suggest to an enemy many more soldiers. The rebel air force was a couple of old P-51s that Bissell had arranged to have the Nicaraguan dictator Anastasio Somoza buy and lease out for cover. When real bombs ran out, the planes dropped Coke bottles that emitted a sinister whistle. This was the perfect symbol of American intervention circa 1954: the Coca-Cola bottle as bomb, bringing with it American-style fear and intimidation.

Success in Guatemala lent the CIA a sense of false confidence that would ultimately lead to the Bay of Pigs disaster.

———

Barely a year later, in the fall of 1955, Allen Dulles called in Bissell, who had been with the agency only briefly, and out of the blue gave him the job of running Aquatone.

Dulles, catlike with his whiskery mustache, tweedy and academic, puffing on his pipe, looked his role. But he was a master bureaucrat. The director had already gotten the Air Force to cover his ass with a letter saying that, indeed, the absurdly optimistic schedule Kelly Johnson proposed for the U-2 was realistic.

While Johnson was building the airplane, the Harvard astronomer James Baker directed the development of special cameras. They would carry 10,000 feet of film on each flight and be able to photograph a swath of territory 125 miles wide and 3,000 miles long.

But they all knew it was only a matter of time before Soviet antiaircraft missiles could reach the maximum altitude of 70,000 feet or so and make the U-2 vulnerable. Even while the U-2 was being tested, Bissell planned the plane that became the Blackbird to fly higher and faster, and in 1958 he began to develop the first spy satellite, Corona.

———

Work on the new plane continued in Lockheed's Building 82 at the Burbank Airport. The first flight was scheduled for August 1955. Kelly Johnson knew who he wanted to fly it first—Tony LeVier, his top test pilot. But first he had to find a place from which to fly it.

He knew they could not test the U-2 at Edwards. By 1954, that base had become too public, and this was a project so secret that its treasury was Johnson's own home mailbox, in Encino, and its cover firm called C&J Inc. (from Clarence Johnson, his given name).

One day in the spring of 1955, Johnson and LeVier set out in the company's Beech Bonanza to look for a test site. To disguise their purpose, they wore hunting clothes and took off in the direction of Mexico carrying a huge lunch LeVier's wife had prepared. They crisscrossed dozens of little airstrips and disused bases. Then finally Osmond Ritland, the CIA's military aide to the program, remembered a strip in the gunnery range where he had been stationed during the war in the desert north of Nellis Air Force Base. With the map spread over Johnson's lap, they aimed for the little x, north of the vast Nevada Test Site. "We looked at that lake," Ritland would later recall, "and we all looked at each other. It was another Edwards, so we wheeled around, landed on that lake, taxied up to one end of it, and Kelly Johnson said, 'We'll build it right there, that's the hangar. We'll put the runway there.' "

The lake was covered with sagebrush. Wild burros occasionally ventured across it. With an old Air Force compass in hand, kicking away the spent .50-caliber shell casings, Johnson laid out the strip. "This will be the tower, right here," he said. Pebbles the size of peas blew around in the afternoon winds.

Soon, seventy-five people would be working here, paving the runway, building hangars, and setting up mobile homes bought from the Navy as barracks. By the time training started, the number of workers jumped to 250. The U-2s were flown in on cargo planes, their wings removed. An official CIA history rather prissily explains, "The site at first afforded few of the necessities and none of the amenities of life."

They flew Bissell out to the site. "Sweet Jesus," Mr. B. may have exclaimed, a favorite phrase of his.

"This will do nicely," he commented. He even liked Johnson's proposed name for the place: "Paradise Ranch."

Johnson and Bissell worked it out such that the Nevada Test Site took official ownership of the strip. Construction on the runway began and a press release was issued, tying the work to the test site, when work began in August. It was done by REECO, Reynolds Electric, the subsidiary of EG&G that ran the nuclear test site. It was referred to now as "Watertown Strip"—perhaps a coy reference to the dryness of the place, or to Allen Dulles's hometown in northern New York State—and not Groom Lake in the emerging cover story, which described the airplane as a weather craft.

Meanwhile, a continent away, the command post for the program was set up in the E-ring of the Pentagon. Bissell moved out of the CIA headquarters on the Mall and into a special program management office with a staff of 225 in an old office on L Street—one of the temporary wartime buildings that had never been removed, an apt metaphor for the survival of the wartime mentality into the uneasy peace. The place was named "Bissell Center," and some in the agency began talking of the RBAF—"Richard Bissell Air Force."

The photo-processing center was set up in a seedy part of town at K and Fifth streets, NW, on four floors above offices of the Stuart Motor Car Company, an auto repair shop. It was code-named Automat.

———

Aircraft ferried workers and materials between the Skunk Works in Burbank and another factory, set up in the little town of Oildale near Bakersfield, a scruffy cotton and oil town where the country singer Merle Haggard had grown up in an old boxcar. The pilots flying to the new secret base were not told where it was. They were simply ordered to fly to a set of coordinates in the middle of the desert and then to await instructions from an unknown air control center, called Sage Control, for further instructions from "Delta." At the point when the radar picked them up, the crews were ordered to descend into the dark desert and lower their gears and flaps. Only then did the runway lights flicker on beneath them.

Between flights, those working at the base lived four to a trailer and could contact families only if absolutely necessary. The phone for this purpose was called the "hello" phone, because that was the only way it was to be answered. It became a fixture of black projects. The number was given out for use only in emergencies. A message was left and the worker or engineer would call back.

There was much drinking and poker. With lots of idle time on their hands, one group of workers fired off homemade rockets made of sawdust, gunpowder, and cigar tubes. Once they nearly hit a cargo plane.

On November 17, 1955, a C-54 making the run from Burbank mistook its altitude and struck Mount Charleston, northwest of Las Vegas, just thirty feet short of its peak. It took three days for a rescue party to reach the crash; an Air Force colonel picked through the wreckage removing briefcases with classified documents. The Skunk Works was lucky; some of its key people had missed the flight because of overindulgence at a beer bash the night before.

———

Curtis LeMay didn't like the idea of a bunch of civilians running an airplane program. But Eisenhower felt with equal certainty that he needed a less biased source of intelligence than the Air Force, which had a record of exaggerating the threat to keep its bomber budgets generous. Protecting himself and the American taxpayer from the military was as important a function of the U-2 program as was protecting us from the Soviet Union. LeMay's deputy, Tom Powers, was flown to the Watertown camp in 1955 and briefed. In the deal that was worked out, SAC would "sheep-dip" the pilots—moving them from military to civilian status and training them. The base now had Air Force and CIA co-commanders.

LeMay carefully planned on letting the agency build the U-2 but then to take it away on behalf of SAC. The Skunk Works and the agency, however, worked to build their own credibility and went over LeMay's head. In December 1955, Secretary of Defense Charles "Engine Charlie" Wilson was flown to the site to bolster his enthusiasm for the program. He talked from the tower to a U-2 pilot high above. Later, Allen Dulles, pipe and all, dropped in to chat with the pilots-in-training.

The Atomic Energy Commission covered the construction work with a brief statement about the building of the airstrip, suggesting it was for nuclear testing activities, and later the familiar weather research cover story was put to work again. On May 7, 1956, a press release was issued over the name of NACA director Hugh Dryden announcing that the new weather research plane had been developed and flown. "The first data, covering conditions in the Rocky Mountain area, are being obtained from flights from Watertown Strip, Nevada." This fooled few people; the Soviets had a copy of it when they shot down U-2 pilot Gary Powers.

At the same time, a long-planned press visit to the X-15 rocket plane at Edwards was hastily expanded to include a look at a "NACA" U-2, which had to be moved from the secretive North Base section of the flight test complex and painted up; it was given a bogus tail number. The paint wasn't even dry when the reporters entered the hangar, and the ground crew was terrified one of the reporters would get close enough to touch the plane. Photos of the weather U-2 look retouched, with the NACA initials in a band on the tail.

———

At four-thirty on the morning of July 14, 1955, the U-2 was loaded on a C-124 and flown to Nevada. By August 4, it had been assembled and was ready to fly.

Test pilot Tony LeVier, who had chosen as his code name for the project "Anthony Evans," was forty-two, at the top of a career wringing out the P-38 for Lockheed and then flying the first Skunk Works air-plane, the XP-80 jet. He was fourteen when Charles Lindbergh flew the Atlantic, and he'd immediately begun earning money collecting old tires and other junk in his Whittier, California, neighborhood, paying five dollars for his first airplane ride. Beginning with the Waco 10, in which he soloed three years later, LeVier would fly more than 250 different air-craft. By the time he came to Lockheed in 1941, he was already well known as a stunt and aerobatics pilot. He flew such exotic craft as the Mendenhall Special from Muroc Dry Lake in 1936 and won major tro-phy races in 1938 in a craft called the Schoenfeldt Firecracker.

At Lockheed, he immediately helped to figure out the odd high-speed compression problem of the P-38—a precursor of the shock waves at the sound barrier—and would put in more hours in its cockpit than any other test pilot. In June 1944 he made the first flight in the XP-80, and then flew its successor, the XP-80A "Gray Ghost"—of all the planes he flew, the one that he said came closest to killing him.

In March 1945, LeVier pushed the jet past 550 miles per hour when a turbine blade let go and he found himself embarrassed by the sudden lack of a tail. The plane began to tumble, and with the G's he could barely reach the canopy release handle. When he did, it came off in his hand. Reaching behind the seat he grabbed the raw cable. Finally, the plane turned over and dumped him out and he pulled himself up into a little ball, waiting for what was left of the airplane to strike him. At just 3,000 feet he managed to get his parachute open.

Yet for all the near escapes and the flamboyance, LeVier developed into the most scientific and cautious of test pilots. He was not a wild-eyed Yeager type, but was obsessed with safety. He had seen too many guys go in. In his retirement he would establish an organization to teach better, safer flying practices and was constantly frustrated with the lack of support he got from government and industry. He developed such practical and basic safety devices as the master warning light system, the trim switch on the control stick, and the afterburner igniter.

With the U-2, LeVier would take no more risks than necessary. It was hard to see out of the cockpit and hard to get a sense of horizon; he wanted the landing strip painted with markings, but the penny-pinching Kelly Johnson found the four-hundred-dollar expenditure excessive. Finally LeVier himself put strips of black electrical tape on the canopy to indicate the true horizon.

———

U-2—"Utility 2"—was the innocuous and noncommittal tag for the plane. But another story circulated about the source of the name.

The plane's long wings gave it so much lift that it was hard to land. The first flight happened by accident: LeVier took it out for a taxi test, but the airplane took off. "It went up like a homesick angel," LeVier said later, more for quotation than anything else. "It flies like a baby buggy." The only problem was it didn't want to come down. In the C-47 chase plane, Johnson kept after LeVier to land nose down, but the plane kept porpoising—it would get down into ground effect, the area where the proximity of the ground added to its lift, and begin a forward and aft wiggle, the "porpoise." After five or six tries, and mounting tempers on both sides, LeVier came in and did it the way he wanted to in the first place—he stalled the plane to get it on the ground.

Once they were both down, Johnson and LeVier continued to argue. "What the hell were you trying to do, kill me?" LeVier said. He gave Johnson the finger. "Well, fuck you."

"And fuck you, too," Johnson replied.

The "you, too" attached itself to the airplane. Or so the tale goes.

Within minutes of the landing, a heavy rain began, the first in months, the equivalent of the lake's total annual rainfall.

That night there was a big beer bash and the arm wrestling that Kelly, proud of the arm strength he had acquired putting up the wall laths during his youth, always fostered. "You did a great job," he told LeVier, calmed down now. When they arm-wrestled, Johnson took LeVier down right away.

The next morning, LeVier appeared with his arm bandaged, wanting to make the point that Johnson had injured his chief test pilot. But Johnson remembered nothing of the night before.

———

The project was variously termed "Aquatone" or "Idealist," but for a long time the plane was just referred to as "the Article," as in the military phrase "test article." Soon some of the Skunk Works folks were referring to it as "Kelly's Angel."

After a character in Milton Caniff's comic strip *Terry and the Pirates*, it was later nicknamed "Dragon Lady." Terry and the Dragon Lady were erstwhile enemies who had become tenuous friends as the Cold War brought hostility between Taiwan and mainland China. The reference suggested the uneasy relationship between the pilot and the tricky airplane that was the triumph of Kelly Johnson's Skunk Works—the first example of the new kinds of weapons the Cold War demanded.

———

To take off, the U-2 wore long, drop-off wheels on its wing tips—"pogo sticks," they were called—and one pilot said they made the aircraft look like a vulture on crutches. That was the right image: The U-2 was delicate and shifty to fly. It would kill several men at the Ranch before it ever went overseas.

With a skin just $\frac{2}{100}$ of an inch thick, the plane's aerodynamics left only a tiny window between overspeeding and stalling. Its fuel could shift suddenly and throw it off balance, its engines were prone to flaming out, and its wings were so long they could snap with sudden maneuvers.

After the training operation was moved from the Ranch to Laughlin Air Force Base in Del Rio, Texas, in June 1957, one eager young pilot decided to fly his plane over his house to show off; he banked, dipped his wings, then stalled and crashed. He died in front of his family.

In early 1956, a pilot suffered a flameout over Tennessee and radioed back. Using a procedure set in place, where sealed envelopes had been left at selected SAC bases for just this eventuality, Bissell had the pilot directed to Kirtland AFB in Albuquerque. Then Bissell phoned the base commander and told him that in about forty-five minutes a secret plane would be landing at his field and he should immediately cover it and phone for further instructions. A half hour later Bissell got word that all had gone as planned. The U-2 had that much glide range.

———

What Frank Powers remembered about Watertown, as he knew the airstrip at Dreamland, was the food. There wasn't much to do—a movie at night, a couple of pool tables, no bar, no club. He looked forward to getting back to Burbank on the weekends and the return from his new identity: from "Francis G. Palmer" back to Francis Gary Powers.

But the food was excellent. It was better food than in Turkey, where Powers was to be stationed—better food, to be sure, than in Lubyanka or Vladimir prisons, where he would be faced with fish soup, endless rations of potatoes, and, once a week, the highlight of the fare, a cube of meat the size of a thumbnail.

After his U-2 was shot down on May 1, 1960, Powers thought back to the food at the Ranch, as he called it, just as he had learned to refer to the CIA as "the company," or "the government." Before his release, he would lie on his cot, dreaming almost nightly of banana splits and coconut cream pies, hamburgers and green salads. Once, he argued with his cellmates about whether we dream in color or only in black-and-white. He resolved the argument that night. He had a dream that he clearly remembered was in color—a banquet of food and wine. Before he could taste any of it, he woke up.

————

Powers came to the Ranch in the spring of 1956, in the second class of pilots to be trained to fly the U-2. The week before, another pilot had bought it in the first U-2 to crash. In September 1956, Howard Carey, a pilot friend of Powers's from the Ranch, was killed in Europe after a couple of curious Canadian interceptors zoomed by his U-2 for a closer look. The wake of the fighters tore the spy plane apart.

Powers had been excited about the boldness and daring of the U-2 scheme from the moment he had heard of it. Like many, he felt that the United States had stalled the Cold War after "the stalemate and compromise in Korea." He already had a top-secret clearance: At Sandia Air Base in New Mexico, in 1953, he had gone through training for delivering nuclear weapons.

At the Ranch, he noted the miles of uninhabited land surrounding the little strip; in the airplane, which needed only a thousand feet of runway to soar into the sky, he enjoyed a feeling that he called a special aloneness.

————

The Skunk Works would cite the numbers forever after: It had taken just eighty-eight days to produce a prototype, eight months to fly the first plane, and now eighteen months to provide an operational spy craft. Overflights of the Soviet Union began in July 1956. The first go-ahead was for just ten days of flying. Eisenhower was leery, despite the promise of Bissell and others that Soviet missiles would never reach the U-2.

Soon it was clear that the whole thing had paid off. The pilots looked for bombers and missiles, tracked nuclear tests with filter paper that recorded the products of the explosions, even flew through clouds of fallout. They monitored and recorded radar and telemetry frequencies, and they actually learned a lot about the weather over the Soviet Union, their cover story.

When the first photos came back, Eisenhower and Dulles spread them on the floor of the Oval Office and looked at them with glee. The airplane discovered untold intelligence riches. In July 1955, the Soviets had shown off a mass of new bombers at their annual Aviation Day parade, and the bomber gap was born. Now, one U-2 pilot had found a base with thirty Bison bombers on the tarmac—was this evidence of a major buildup? Additional flights showed that this was the only base where the Bison bombers were stationed; it was the entire fleet. The bomber gap closed. Eisenhower was able to keep Curt LeMay's demands for more B-52s and the B-70 in check. Richard Bissell's friend the columnist Joe Alsop would later leak word of the operation to the public.

Another flight revealed the space facility at Tyuratum, the Cape Canaveral of Russia. A third flight located a tower that looked like a nuclear test facility. The CIA scoffed, but two days later an explosion was recorded at the previously unknown facility. Additionally, the U-2 found evidence of new radar facilities that made it—and successor airplanes—even more vulnerable to detection.

The U-2's most secret flights, however, were not over Soviet airspace. They were the ones that spied on the English and French and Israelis, beginning with the Suez Crisis in 1956. From them, Eisenhower learned that the French and the Israelis had lied to him—that they had many more Mirage fighters than they had acknowledged. And after the fighting began, one U-2 did two passes over Cairo West airport in a couple of hours, capturing the before and after images of a bombing attack.

Officials had figured on getting two years out of the U-2. By 1960, they had gotten four. But the Soviets were tracking the flights on radar, as they had been almost from the beginning, and their surface-to-air missiles (SAMs) were getting closer. The president often ordered the flight plans changed or the flights delayed. It drove the agency and the Skunk Works people crazy. They called Ike "Speedy Gonzales."

The last flight, called "Grand Slam" because it would fly all the way across the Soviet Union, south to north, was approved for late April.

Weather delayed the flight. The unit shipped from Turkey to Pakistan, where the flights operated from temporary setups. The scheduled airplane, which had the best record, turned out to be due for maintenance; instead, Powers got number 360, a known "dog." (The planes were basically built by hand and tended to have individual differences and eccentricities. Some were sturdy performers, others plagued with gremlins. Flying out of Atsugi, the Japanese U-2 base, 360 had once made an embarrassingly public crash-landing on a muddy airstrip, where armed guards chased off a crowd of camera-toting Japanese.)

The plane was constantly developing new and different technical maladies. During Powers's flight, the autopilot quickly began to go on the blink. Tracking Powers, the Soviet military launched a salvo of SAMs. Nine miles above the earth, Powers was writing in his logbook when he saw an orange flash.

His first thought was "I'm done for." Then the wings went and the fuselage began spinning. Powers's legs were pinned against the panel by the force. He couldn't eject, or his legs would be taken off above the knee. He decided to scramble out of the cockpit but found himself held in by his oxygen lines. He tried to reach the destruct button. He got within six inches of it. Then he decided he had to try to save himself. He got free of the plane. Floating beneath his parachute, he saw rolling hills, a forest, a lake, a village. In its early spring greenery, it reminded him of his native West Virginia. He remembered a map in his pocket showing alternate routes back to Pakistan and Turkey. First taking off his gloves, he pulled out the map and carefully ripped it into little pieces and scattered them. Then he thought of the coin and the poison pin inside: a silver dollar with a hidden pin laced with curare—a device for suicide.

It was the first time Powers had decided to carry the silver dollar. He did so on a whim, thinking of it vaguely as a potential weapon, not a means of self-destruction.

Then a sense of the absurdity of the device replaced his previous admiration for its cleverness. What better token of a capitalist spy pilot than a silver dollar? It was just the sort of James Bond gadget that people expected the CIA to come up with—and the agency had tried to meet their expectations. Who in 1960 used silver dollars anymore, except on ceremonial occasions?

Powers pulled the pin from the coin, hid it in a pocket of his flight suit, then dropped the silver dollar.

He saw a second parachute blossom above him, which confused him. It appeared that a Soviet pilot had had to bail out too.

On the ground, someone handed him a filter cigarette—Laika

brand, named after the dog who rode into orbit on Sputnik II. It tasted like the Kents he carried in his flight suit pocket.

Twice, the pin escaped discovery in body searches. When the Soviets took his flight suit, though, he warned them about the pin. They tested it on a dog. The dog's tongue turned blue, and it collapsed on its side. Within ninety seconds it stopped breathing; in three minutes it was dead.

He found his interrogators frequently incompetent. There was none of the torture or Korean War–style brainwashing he had worried about. There was much danger, he thought, in overestimating your enemy.

He told the Russians plainly that he had trained at the Ranch, Watertown, strip. His captors came in bearing a map and asked him to point out the Ranch, "to see if he was telling the truth." He pointed to a spot but did not mention that it was a map of Arizona, not Nevada.

_____

The regret in Washington was that the man they had carefully and expensively trained in Dreamland had had the temerity to survive.

Khrushchev fooled Eisenhower with incomplete statements. He hid the fact that Powers was alive until Ike came out with the cover story about a weather flight. The Russians displayed the wreckage of what they said was the U-2, but Kelly Johnson took one look at it and knew they were lying. It was another game, although he never understood why the Soviets had done it. The real wreckage was later displayed in Gorky Park.

It was a classic Cold War mind game. The United States kept insisting that Powers had had a flameout and had descended to a lower altitude to restart his engine, while Powers kept insisting he had been shot down at 68,000 feet, which he gave as the maximum ceiling for the plane. The government wanted to keep the maximum height from the Soviets; Powers wanted to signal his employers not to send over any other pilots, that the Soviets had indeed figured out how to reach the U-2's operating altitude with SAMs. In citing 68,000 as the maximum altitude, which was not true, he was also subtly signaling that he had not told the Russians the real figure.

The Pentagon also wanted to hide the U-2's true operating ceiling in order to preserve public trust in the strength of our nuclear deterrent. How long would it take the press to tell the public that if missiles could reach spy planes above 60,000 feet they could also reach Curtis LeMay's bombers at their lower altitudes? It was a game something like the bomber gap game with the Russians: The president could not reveal that the U-2 had debunked the gap, for which candidate John F. Kennedy was attacking his administration, without revealing the existence of the spy plane.

Forced to chose between admitting he didn't know what was happening in his own administration and admitting responsibility for the intrusion, Ike chose the latter, justifying the need for overflights because the Soviets had rejected his Open Skies proposal, and explicitly citing the danger of "another Pearl Harbor."

In 1962 Powers was exchanged for Rudolph Abel, whom the CIA had described as "a master spy" but who later said he got 90 percent of his intelligence from The New York Times and Scientific American.

The exchange took place on a green bridge between Potsdam and Berlin, a scene out of a John le Carré novel. In a heavy coat and Russian-style fur hat, Powers came into view flanked by a pair of goons, then walked past the thin-faced Abel without acknowledging him.

On the plane home—one of Kelly Johnson's Superconstellations—Powers ate a fine meal of steak and potatoes, as good as anything back at the Ranch.

He was debriefed in a safe house in the Maryland countryside. "What happened to my airplane?" Kelly Johnson asked him. He believed Powers's story and, after the grilling at the congressional hearings, hired Powers as a test pilot flying U-2s. Apparently, Powers never knew his salary was paid by the CIA.

Powers's book came out in 1970, for the tenth anniversary of the flight, and around the same time Lockheed let him go. He then became one of the first traffic-helicopter pilots in Los Angeles.

————

The great national and political coming to terms with the shootdown followed. The planned superpower summit was bust; Eisenhower left office diminished in prestige. The whole incident became surrounded by a cloud of suspicion. The mission had been delayed by the Oval Office, and when the go-ahead finally came, there were problems with the radio and the word had been transmitted by an open telephone land line—a violation of security. Then there was "the Granger," the radar spoofer that the Skunk Works had come up with to fool Soviet radars. If the

Russians knew how the Granger worked, they could have used it as a tracking device. Three Taiwanese U-2s were later knocked down over the People's Republic in a single day in this way.

There was one other dark possibility that Powers wondered about much later. A young Marine assigned to the radar facilities of the Japanese U-2 had defected to Russia in 1959. A formal U.S. government investigation discovered that on three occasions the Marine had spoken to the Soviets of the vital information he could bring with him if they welcomed him. That government investigation was the Warren Commission Report, and the young Marine was Lee Harvey Oswald.

Powers died in August 1977, when his traffic helicopter crashed just three miles from the Skunk Works. He had run out of fuel, but there are those who believe it was no accident.

———

Within months after Powers was shot down, Richard Bissell had the temerity to suggest the program continue flights over the Soviet Union. The president ruled it out. Never again, the country seemed to collectively resolve, would manned spy planes make the pilot and the country that vulnerable.

But in August 1960, the very day Frank Powers stood in the dock in Moscow for sentencing, another of Bissell's secret projects had finally begun to pay off. After more than a dozen failures, the engineers running the Corona spy satellite program successfully recovered a film pod ejected from the satellite whose public identity was Discoverer XIV.

Snatched by a C-119 Flying Boxcar at 8,500 feet, the capsule contained film of a million square miles of the Soviet Union—more than all the U-2 flights together had produced.

This was the future: no human at risk, no violation of airspace. To celebrate, the engineers got drunk and threw one another into a swimming pool. The recovery was announced; the public would not learn of the true spy mission for another thirty years or so.

The most important days of the spy plane, and especially the U-2, were still ahead. In October 1960, Eisenhower got to see for the first time the airplane that had caused him so much trouble, when he stopped in Texas after a trip to meet with the president of Mexico. That same month he approved U-2 flights over Cuba, where the new government of Fidel Castro was showing increasing belligerence toward the United States.

In August 1962, U-2 photos of Cuba showed a shape that photo-interpreters recognized from the thousands of images they had of the Soviet Union: the star-shaped emplacements of Soviet SAM sites, holding missiles like the one that had brought down Powers. In the next few weeks, comparing the new pictures with an extensive database of older ones of the Cuban landscape, they saw more and more sites under construction, and by October they had matched boxes and equipment, carefully measured by computer, with shapes and sizes known from Soviet weapons displayed in Red Square parades: MiG-21s and Sandal missiles. The agency's top "crateologists," experts in all sorts of weapons and equipment packaging, were consulted. It was soon clear that the medium-range missiles—missiles that normally carried nuclear warheads—were being installed in Cuba.

On Saturday, October 12, 1962, Maj. Richard Heyster took off from Edwards North Base in a U-2. He reached the coast of Cuba early the next morning and returned with the key photos showing the six-sided star of SAM sites protecting the medium-range missiles NATO had code-named Sandal at San Cristobal.

When Heyster's take provided Art Lundahl, the head of the photo interpretation office that handled the U-2 photos, with unmistakable evidence of the presence of Soviet missiles, Lundahl hurried to the White House. By noon on Tuesday, he was displaying the photos to the president and his top advisers; a week later, President Kennedy sat in front of the television cameras, announcing the quarantine (the term was chosen instead of "blockade," an act of war in international law).

While the president was speaking, fifty-four of LeMay's SAC bombers joined the dozen that were constantly orbiting on alert. Before the crisis was over, SAC would go from the normal Defense Condition Five to DefCon Two—the highest ever reached. Three days later, Adlai Stevenson, accompanied by staff from the National Photo Interpretation Center, was displaying the wares of the U-2 at the United Nations.

Kennedy ordered more thorough photography of the island, which required low-level, high-speed RF 101 Voodoos—their snouted shadows show up in the most famous treetop close-ups of the shrouded missiles and launchers.

On October 27, Maj. Rudolph Anderson was shot down in his U-2—the sole casualty of the Cuban Missile Crisis, save for several crews of military aircraft that crashed during the mobilization. Anderson's death came just as the Russians agreed to remove their missiles; it was the act, the Soviets said years later, of a trigger-happy local SAM commander.

Even more dangerous was the U-2 that went off course during the crisis and strayed into Soviet airspace near Siberia. "There's always some poor son of a bitch who doesn't get the message," Kennedy remarked with a sigh. Khrushchev had rightly protested that in the current state of tension no one could be sure the spy plane had not been a bomber, the first shot of a nuclear war.

LeMay, the commander of the Cuban reconnaissance group, and Major Heyster were called to the Oval Office for commendation. A photo shows Heyster squeezed on a couch between the bigger officers. "Let me do the talking," LeMay said. But days later, what LeMay talked about was how he had lost. He harangued JFK about how he could have forced out not only the Soviet missiles but the Soviets and Castro as well. We had the Russian bear in a trap, he said, and "we should have taken his whole leg off. Hell, we should have taken his testicles off, too."

———

For decades, the U-2 would continue to be a vital source of some of the most important, detailed political intelligence.

The U-2 victories that did the most to prevent World War III, however, were the ones over President Eisenhower's military-industrial complex. Such intelligence provided support for those resisting the building of more bombers, more missiles. The numbers that did get built were huge, of course, but without the solid information to counterbalance the Curtis LeMays, they would have been far greater and the temptation to use them much stronger.

11. The Blackbirds

The U-2 may have been "Kelly's Angel," but even before it had flown over the Soviet Union, it was clear that what the CIA needed was an "Archangel." Radars and missiles were improving. From the very first U-2 flight, CIA operatives and Skunk Works technicians had been surprised at how quickly the Soviets learned to track the craft on radar. The U-2 would have to be replaced.

Richard Bissell had the Skunk Works look into the best ways to escape detection by radar—speed, height, reduced radar profile, or some combination of the three. The result was the most heroic story in the whole Skunk Works buffs' catalog.

The Skunk Works plunged into an extensive study of a superplane powered by hydrogen. Project Suntan, as it was called, cost taxpayers the equivalent of two billion of today's dollars before officials realized that creating a whole system of refrigerated tanks and pipes for liquid hydrogen at bases around the world would cost billions more. The program would remain secret for nearly twenty years.

In 1960, the CIA finally settled on a Skunk Works plan for a high-flying conventionally powered craft, flying so high and fast—three times the speed of sound—that it could elude missiles and fighters. It would be built of titanium, the first such use of the metal. The program was called Oxcart. Eisenhower, increasingly apprehensive about the U-2, just called it "the big one."

Secrecy was even more intense, if that was possible, with Oxcart than with the U-2. Checks were made out to the dummy C&J corporation. Drawn on the CIA's reserve funds, free from overzealous congressional or executive auditing, they were sent to anonymous post office boxes scattered around Los Angeles. Once, a suspicious supplier tried to track down the box; he was intercepted by security agents.

The Ranch, Watertown—or "home plate," as some were now calling it—prepared for a much larger effort than the U-2 had required. The new plane, called the A-12, or Blackbird, would need a longer runway and larger support staff.

Construction began in earnest in September 1960, and continued on a double-shift schedule until mid-1964. The new runway measured 8,500 feet and required pouring over 25,000 yards of concrete. Kelly Johnson was concerned that with the high takeoff speed of the Blackbird, expansion joints could set off dangerous vibrations, so the runway was built of offset slabs, each 150 feet long and layered in tile-like patterns.

The Blackbird would also need about 500,000 gallons of PF-1 aircraft fuel per month. After considering an airlift or a pipeline, the team decided to rely on trucks, but that required paving eighteen miles of highway leading into the base.

SAC again provided support. In late 1961, Air Force colonel Robert J. Holbury became commander of the base, with a CIA manager as his deputy. Support aircraft began arriving in the spring of 1962—including eight F-101s, two T33s, a C-130 for cargo transport, a U-3A for administration purposes, a helicopter for search and rescue, a Cessna-180, and a Lockheed F-104 for chase. The Blackbirds were too big to be loaded onto planes and flown in from Burbank, like the U-2s had been, so they were carted by truck. A pilot truck outfitted with thirty-five-foot bamboo outriggers—the size of the finished airplane—drove the route testing for obstacles, such as signs, branches, and so on. Then the obstacles had to be removed, sometimes through negotiation with local authorities. Road signs were hacksawed and hinged for the passage of the new bird.

Between the high-tech complexities of working with titanium and the lower-tech problems—they tested the ejection seat by towing it with a 1961 Ford Thunderbird convertible, the fastest car they could rent from Hertz—the Skunk Works fell behind schedule on the Blackbird's first flight, and there were warnings from Richard Bissell.

For the first time, the Skunk Works was falling behind. The initial flight was originally planned for the end of May 1961, but it slipped to August, largely because of Lockheed's difficulties in procuring and fabricating titanium. Ironically, much of the raw metal would come from the Soviet Union.

Not surprisingly, the manufacturer of the engines, Pratt & Whitney, found it difficult to turn out a power plant to drive the big airplane to three and a half times the speed of sound.

It must have galled Johnson to admit the delay when he got a stern note from Bissell:

I have learned of your expected additional delay in first flight from 30 August to 1 December 1961. This news is extremely shocking on top of our previous slippage from May to August and my understanding as of our meeting 19 December that the titanium extrusion problems were essentially overcome. I trust this is the last of such disappointments short of a severe earthquake in Burbank.

But delays could come as no surprise, since the Skunk Works was single-handedly pioneering the use of titanium, learning on the job that the metal had to be carefully protected against contact with chlorine, fluorine, and cadmium, which could make tools unusable. The engineers had discovered that the Burbank water supply was fluoridated, and from then on used only distilled water. A whole new family of lubricants and seals had to be invented, and even so, the airplane literally seeped fuel when it sat on the ground with full tanks. For its whole flying life, the Blackbird had to be "topped off" by in-flight refueling once it was in the air and expansion had tightened the tanks.

Kelly Johnson predicted that the craft would not be matched for the rest of the century. It was like a piece of technology retrieved from the future.

————

In January 1962, an agreement was reached with the Federal Aviation Administration that extended the restricted airspace around the test area. The first references to Area 51 began to appear around this time, as well as a new name applied to the control tower for the airspace: Dreamland.

A number of FAA air traffic controllers were cleared for the project, and the North American Air Defense Command (NORAD) established procedures to prevent their radar stations from reporting the appearance of the Blackbirds on their radar screens. But on the high radar range at Tonopah, operators would soon be seeing things moving much faster than they could explain.

On February 17, construction of the first aircraft was finished, and in the next few days the plane underwent its final tests. It was taken apart and stowed on the special trucks designed to move it to Groom Lake. A famous film clip shows Kelly Johnson planning the movement of the first A-12 to the base. On the chalkboard behind him is this list:

Feb 17	Aircraft complete
Feb 18	Aircraft put down on its gear
Feb 19–Feb 22	Engineering final tests
Feb 23–25	Disassemble and load on trucks
Feb 26	4:00 AM—Move out to Area 51

On February 26, 1962, at two-thirty in the morning, the convoy bearing the first plane left Burbank. It arrived by the back road to Groom Lake at about one in the afternoon on February 28. The second aircraft struck a Greyhound bus en route; the bus company was quickly and quietly compensated some $4,800 to settle the damage.

Not until April was the plane ready to fly. On April 26, 1962, pilot Lou Schalk flew the plane for about a mile and a half, just twenty feet off the ground. The plane felt like it was wallowing, and he decided to set it back down. From the ground, the crew saw the plane begin a series of lateral oscillations, which terrified Johnson, who later recorded that "it was a horrible sight."

The tower could no longer hear Schalk, and from the tower and the ground you could see the Blackbird disappear in a great cloud of dust from the lake. It was enveloped for minutes, then finally reemerged in the distance as Schalk made a turn. They were relieved he hadn't run into the mountains.

The first "official" flight took place on April 30, a year behind schedule, and on the second flight, on May 4, the plane went supersonic. One spectator at the first flight was Richard Bissell. He had been eased out of the CIA in February 1962, a dismissal occasioned by the Bay of Pigs fiasco the previous April and delayed only for the sake of appearances. At Kelly Johnson's personal invitation, Bissell was standing on the white surface at Groom when the long bird he had championed took off.

Space was the next frontier of espionage, and the first flight of the

Blackbird coincided almost exactly with the orbiting of the first Soviet spy satellite. From now on, airplanes at Groom would have to be kept in hangars or covered with camouflage when Soviet satellites passed overhead, as they would be sure to do. In the Kremlin, they already knew about Dreamland.

————

The pressure to get the Blackbird operational mounted in the fall. In January 1963, Bob Gilliland arrived at the test location, ready to fly the Air Force fighter version of the plane, joining pilots Bill Park and Jim Eastham. On May 24 came the first crash, when the pitot tube iced up and left pilot Ken Collins with no accurate speed indication. He bailed out over Wendover, Utah. A farmer in a pickup truck found him. "I've just crashed an F-105 with a nuclear weapon on board," Collins said. "Let's get out of here and find a phone." The farmer quickly complied.

On August 7, 1962, the AF-12, the fighter version of the Blackbird, first flew. A whole family of Blackbirds was hatching in the desert, and they could not be kept hidden much longer.

Shortly after he became president, Lyndon Johnson, briefed about the Blackbirds, ordered that preparations be made to reveal their existence. It was an election year and crucial for the president to appear tough on defense. The leading Republican candidate, Barry Goldwater, had already begun to criticize administration defense policy.

At a press conference on February 24, 1964, Johnson read a statement that described the new "A-11" as "an advanced experimental jet aircraft." (For some reason, Johnson said A-11 instead of A-12. Similarly, when he announced the SAC version of the Blackbird, Johnson misstated the assigned name—RS-71, for "reconnaissance strike," instead of SR-71. The brass and contractor scrambled to invent "Strategic Reconnaissance" to back up the reversed initials.)

There was no mention of the first Blackbird, the CIA spy version. And, of course, there were no "A-11s"—the Lockheed design number for the fighter version of the Blackbird—at Edwards, so some were quickly flown from Dreamland to the base. As the "Oxcart Story," the official CIA history of the project, reported, "So rushed was this operation, so speedily were the aircraft put into hangars upon arrival, that heat from them activated the hangar sprinkler system, dousing the reception team which awaited them."

In July 1964, pilot Bill Park nearly lost his life when a servo locked up and set his plane rolling just five hundred feet above the runway. In December of the next year, Mele Vojvodich ejected safely at an altitude of 150 feet on takeoff: An electrician had reversed the yaw and pitch gyros—in effect flipped the controls—and the result was another fireball on the lakebed.

In November 1964, the airplane was pronounced ready for use. As early as October 1962, the agency had been eager to offer the still-adolescent Blackbird for spying on Cuba, where the U-2s were vulnerable to SAMs. And in the fall of 1964, Khrushchev threatened to shoot down U-2s over Cuba after the presidential election. At Dreamland, hasty preparations were made to ready A-12s for the job if the Soviet premier carried out his threat.

———

The Blackbirds were not black at first, but metallic, except in front of the canopy and on the edges, where they were painted dark, like the grease-paint on a football player's cheekbones to cut the glare. Now Ben Rich had the idea to paint them black, to deal with the heat of high-speed flight.

To showcase the new plane's abilities, on December 21, 1966, pilot Bill Park flew 10,198 statute miles in six hours. Taking off from Dreamland, he started north toward Yellowstone National Park, then eastward to Bismarck, North Dakota, and on to Duluth, Minnesota. Turning south, Park passed Atlanta en route to Tampa, Florida, then back northwest to Portland, Oregon, and southeast to Nevada. The flight continued eastward, passing Denver and St. Louis. Turning around at Knoxville, Tennessee, Park slipped by Bob Gilliland's hometown of Memphis in the home stretch back to Nevada. This flight established a record unapproachable by any other aircraft.

———

The guys on the start carts—the big twin Buick and Chevy V-8s they would roll up to "crank" the engines on the Blackbirds—really liked Bill Park. So it was especially tough for them to watch, from the south end of the Groom runway, as the long black plane, just five hundred feet above the lake, went careening to one side and began rolling steadily like a boat going over, until they could see only the bottom of the airplane, with its landing-gear doors and streaks and smears, finally plunging down. The lake filled with an ugly orange balloon of flame, blackening at its edges.

They were talking on the phone to the guys back at the hangar trying to figure out what the hell had happened when Park walked up, the picture of calm.

―――――

Park hadn't had much time to think. He couldn't get the plane to respond. It wouldn't stop rolling. It was down to two hundred feet when he flicked the arm switch for the seat, then leaned forward and grabbed the big D-ring between his legs—like some big, loopy luggage handle—and leaned back, putting his weight into it. And suddenly the top was gone, the air rushing cleanly through the cockpit, and in another fraction of a second, still in the seat, he was sailing up into the rangy mountains around Dreamland.

Kicking Park up the pole was the rocket engine under his rear end. As soon as it quit, he got another kick as the drogue chute opened, and then the seat ejector, the straps they called "butt snappers," threw him out, the way it was supposed to, but he could see he was pretty close to the ground, and it must have felt like forever before the chute finally opened. Two things seemed to happen at the very same moment. First, he was grabbed by the chest and legs as the chute went taut, drawing him upright. And second, his feet hit the ground. Then he gathered up his chute and began walking toward the end of the runway. It was not the last time Park would eject at the Ranch.

―――――

For the test pilots, Dreamland was just the office, their everyday job. They saw very little that was exciting and certainly nothing mysterious about Groom Lake. Secrecy was a burden, a frustration. But the difference between a test pilot and a regular pilot, said Bob Gilliland, is that test pilots have emergencies every day. What caused the most fear? one Blackbird pilot was asked. Fear? He wouldn't touch the word. "Sure," he said. "From time to time there were *levels of concern.*"

The basic mode of life on the Ranch was akin to the mind-numbing tedium characteristic of military installations the world over. One ground-support man tried to liven things up—and, it must be speculated, supplement his salary—by showing pornographic movies. Blue movies for the blue sky boys! But Kelly Johnson got wind of it and put his foot down, albeit softly. "Whatever you've got up there, I just want it out," he told the man.

———

Bob Gilliland came into the program through his friend pilot Lou Schalk. It came about because of a problem with Bob's Mercedes. Bob had to drop it off at an auto shop on Sunset Boulevard and he got Schalk to pick him up. Schalk had a red Austin Healey, and on the way back the two pilots, jammed together in the little British sports car, began talking. Schalk told Gilliland that he was flying a new airplane Kelly Johnson was developing and that he needed another test pilot.

Gilliland was having fun flying the F-104, a real hot rod of a fighter, and he was afraid the new plane was some weird settle-on-its-butt thing like the vertical-takeoff-and-landing craft the Skunk Works had dreamed up and Herm Salmon had flown. But he agreed to talk to Kelly Johnson. Johnson told him that the new plane "will be faster and go higher and farther than the 104." That got Gilliland interested.

"Now, let's go take a look," Johnson said. He led Gilliland into the hangar where the next Blackbird still lay in long, sharp pieces. Gilliland could sense an excitement in the very shapes, as Johnson knew a good pilot would. He signed on.

———

Johnson hated the military test pilots. He always wanted his own pilots to test the new planes thoroughly before the military boys could get their hands on them. It all went back to 1939, when Ben Kelsey, an Army test pilot, lost the prototype of the P-38 trying to fly it across the country to set a new record. It would impress the brass and Congress and the public. But he came in too low on landing, an engine gulped and hesitated, and he ended up in a bank on a golf course.

It set the program back two years. Tony LeVier went so far as to say that that particular piece of grandstanding prolonged the war. So Johnson picked his test pilots carefully, and their succession is as legendary in the aviation world as that of Yankee centerfielders: Milo Burcham on the P-38, Tony LeVier on the XP-80 jet and F-104 Starfighter, Lou Schalk, Jim Eastham, Bob Gilliland, Bill Park.

Security was intense. Lockheed even airbrushed the mountains out of the background of photographs to help disguise the location. The situation was much changed from the early days of the U-2, the surplus Navy structures supplemented with brand-new modern hangars, and a workforce that had grown by five or six times, to 1,100 or so by 1962.

Life at Groom was dull, but Bob Gilliland would go jogging and lift weights sometimes to shed stress. You could play tennis and there was a softball team, but not much else. Lou Schalk found other diversions: He brought the red Austin Healey to the base and raced it across the lake bed against Jim Eastham's blue one.

But it was exciting learning how to carefully move the inlet spikes, like big missile nose cones, that were the key to engine performance, to make the airplane go higher and faster. It was exciting, Gilliland always felt, because almost every evening meant the end of another day when he had been able to fly faster than any pilot in history. Only no one knew. By the fall of 1963 they were flying the airplane well beyond Mach 3, at 110,000 feet.

There was no television, only radio, at the Ranch, and one day in November 1963 Gilliland came back from flying "the CIA bird," the A-12, to find that everyone in the hangar was gathered around the radio. "What's going on?" he asked. "JFK has been shot, LBJ has been shot, Connally has been shot," someone told him. Bill Park said, "Well, hell, I don't know what all the excitement's about. It's just another Texas shootout." The pilots by then were as dry and hard as the bed of Groom Lake. They seemed to have absorbed the desert itself.

Park was the driest. Ben Rich called him "an outstanding stick man, cool and calm," but there was more. When there was discussion of basing U-2s on aircraft carriers, it was Park who was called on to see if it could work. He landed one on the pitching deck. Using a special technique he devised himself, in 1958 Park had pushed the F-104 to a world record altitude of 91,985 feet—a record the Skunk Works had to keep secret.

When Kelly Johnson strapped the D-21 drone onto the back of the Blackbird, in the program called Tagboard, he picked Park to fly the dangerous release missions. The idea was to send an unmanned craft over China to take a look at the far western Lop Nor nuclear test area.

After takeoff from Groom Lake, the launch run would begin over Dalhart, Texas, aiming for a release point around Point Mugu in California. The Blackbird took half a continent's width just to get warmed up to Mach 3 plus, the speed necessary for the D-21's ramjet to function.

The very first time they got the mother plane up to speed, the D-21 let go all right, but it stuck close to the big airplane as if reluctant to venture off on its own. Then all of a sudden it veered and dropped—the engine gulped and faltered—and hit the tail of the Blackbird, pitching the big plane forward. The long black plane snapped in the middle and, still traveling at Mach 3, began to tumble down, as both Park and Ray Torick, the backseat man in charge of the launch, flipped the levers to arm the ejection seats and pulled the rings. In the Pacific off Point Mugu, Park was picked up by rescue helicopters, but Torick's pressure suit filled with water and dragged him to his death.

Kelly Johnson immediately canceled the D-21 test program; it would remain secret for more than a decade. The strange black D-21s would make their way to Arizona, where years later, in the boneyard, I would see them—or I wouldn't.

————

Park would become the longest-serving pilot at Dreamland. He went on to fly Have Blue, the first Stealth prototype. When he first saw it, Park couldn't imagine how the thing would ever fly. A lot had been sacrificed to get the right radar cross sections. Thinner and more dartlike than the Stealth fighter to come, it was painted up in a desert camou, a broken pattern of grays and browns. It also had what was known as an "an excessive sink rate," a tendency to fall like a stone in certain low speeds. Its wings were too small. This could be fixed in a production fighter, but it was something the test pilots had to live with.

Coming in one day in 1976, the plane took a dip on Park and hit hard on the right gear. He pulled up and around, but when he started to lower the gear again, the right would not go down. He even came down on the left and tried to shake the other gear lose, without success. Park took the airplane up to ten thousand feet and burned off most of his remaining fuel. "Unless anyone has a better idea," he radioed, "I'm bailing out." That morning, the commander had asked him about letting the base paramedic go for the day. There hadn't been any problems on earlier flights, and the test series was nearing its end. But Park demurred. He went by the book.

Now he had to pull the ring again, but he struck his head on the headrest, cracked a vertebra, and was knocked out. Amazingly, the seat lifted him free of the airplane at ten thousand feet, he separated from the seat, and the chute opened as designed. But still unconscious, he hit hard. He broke a leg and his head was dragged along the caliche, where his mouth filled with dry sand. By the time the paramedics got to him, his heart had stopped. He would spend six weeks in intensive care and six months in a cast.

The quintessential Park story is not of any of his bailouts, but of an earlier close call. He was flying a U-2 out of Burbank when it developed fuel problems. The engine quit and he had to dead-stick it home. He barely made it back to base, clearing a chain-link fence by six inches. Ben Rich came out to the airplane. "What happened?" Rich asked. "I don't know," Park said. "I just got here myself."

———

The secret black planes would challenge SAC, and LeMay. Dreamland was, indirectly, an offspring, too, of the blue-sky, high-noon vision that was SAC. While Dreamland would birth black planes, they would serve SAC's silver bombers. They would find the targets for the bombers to strike. And they would, the day after doomsday, fly back to see how well the silver planes had done.

One day in 1962, Richard Bissell came to the White House to brief JFK on the new and still very secret Blackbird, the CIA's A-12. The president was puzzled. He looked at the documents, and listened to Bissell telling him how far and fast the agency's new plane could fly. "Could Kelly Johnson convert your airplane into a bomber?" the president asked. "That question is more properly addressed to General LeMay," Bissell diplomatically answered.

But it got Johnson in hot water, and he was not pleased with Bissell. Johnson had carefully not spoken to the Air Force or the Pentagon about the bomber version of the Blackbird. He knew LeMay wouldn't like a black challenge to his silver planes. But now he hurried to Washington to work his charm on the bypassed generals. Later in 1962, when the B-70 was cut back from ten planned planes to four, LeMay blamed Johnson.

Finally, the two men met at the Skunk Works. Their aides drew back as they walked and conferred, the cigar smoke trailing behind them. By the end of the day, it appeared Johnson's charm had had an effect. LeMay seemed all set to order bombers and recon Blackbirds—an entire black air force. But by the end of the year, only the SR-71—the reconnaissance version—had been ordered.

With the SR-71, SAC got its own Blackbirds, and while it used them to spy on distant countries—the Soviet Union and China excepted—their ostensible job was something called "post-strike reconnaissance." The SR-71s came in handy in 1973, when the United States eased tensions in the Mideast by offering the Soviets photographic proof of Israeli positions. The Soviets had threatened intervention; now they backed off.

Primarily, though, the SR-71s were supposed to be part of SAC's main "deterrent" mission—fighting nuclear wars. The SR-71s' job would be to fly over the Soviet Union in the event of nuclear war, after the bombs and missiles had fallen, and "assess battle damage." What they did do was simple intelligence gathering, flying over trouble spots, and, after the A-12s were phased out in 1968, they did it for the CIA as well as the Air Force.

But to keep up the original premise, SR pilots had to go through the monthly ritual of refresher courses in post-nuclear-battle damage assessment, learning to distinguish what cities were in need of additional blows and on which targets another nuke would simply, in the infamous phrase of the overkill era, "bounce the rubble." The exercise struck the pilots as not only pointless but grim and surreal.

Secretary of Defense Robert McNamara was looking at different kinds of planes. McNamara—"Mack the Knife," the contractors called him—was pushing the THX for both the Air Force and the Navy.

Both the Air Force and the Navy hated it—if only because it forced them to share. McNamara not only killed the Blackbird but put in motion the process that would destroy the tooling to produce it. The Skunk Works buffs all know the dark day: On May 5, 1970, Kelly Johnson was ordered to sell the dies and jigs for the fastest, highest-flying airplane in history as scrap for just a few cents a pound.

But the symbolic end had come on June 26, 1968, when a group of high-level CIA officials, pilots, and pilots' families assembled on the caliche at Groom Lake. Each of the living agency pilots, and the families of those who had died, were presented the agency's highest award, the Intelligence Star. At last wives got a glimpse of the strange place to which their husbands had been disappearing. But there would be no public acknowledgment of the existence of the CIA Blackbird for another two decades.

12. Low Observables

The Blackbird had succeeded because of the great speed and altitude at which it flew. But before the plane ever took off, the Skunk Works people knew it could never safely fly over the Soviet Union. Constantly improving Soviet radars, such as the one the Pentagon called Tall King, could spot it even if the human eye could not.

The Blackbird looked as if it had been shaped for speed, but the knifelike extensions of its fuselage, called chines, which made it look like a sword with wings, had been designed to elude radar. Future airplanes would take their shapes less from the wind tunnel than from radar test chambers and be sculpted not by shock waves but by electromagnetic ones.

————

In the 1860s and '70s, in quiet labs at King's College London and Cambridge, the Scottish physicist James Clerk Maxwell theorized that light, electricity, magnetism, and what would later be called radio- and microwaves were all related. In his 1873 *Treatise on Electricity and Magnetism*, he told how they shared the properties of reflection and refraction, diffraction and polarization. Here lay the origins of radar.

One day in 1932, with fascism taking hold across Europe, former prime minister Stanley Baldwin stood up in the House of Commons to deliver a warning that there was no longer any question of protecting the man in the street from bombing. During World War I, zeppelins had bombed London, and the British understood that no fleet could provide full defense in the future. The next war would turn on the fact that Britain "was no longer an island." "The bomber will always get through" was Baldwin's famous phrase, to be repeated down the decades in the debate over airpower.

"The only defense is in offense," he went on, "which means that you will have to kill more women and children more quickly than the enemy." It was an endorsement of the teachings of the airpower enthusiasts and a foretaste of the doctrines of massive retaliation and assured deterrence.

Not everyone could accept that there was no defense against the bomber, however. Just as Ronald Reagan, a half century later, would look to the dream of Star Wars to break the logjam of mutual destruction, the British air ministry desperately sought new ways to shoot down bombers. They even looked at such exotic ideas as radio-wave weapons—ray guns. How much radio energy would it take to make a man's blood boil? scientists were asked. How much to blow an airplane out of the sky?

The results were not promising, but something else interesting came out of the discussion: the idea that you could locate, if not destroy, an airplane by beaming radio waves at it and capturing and measuring the reflection. Clerk Maxwell had postulated in 1873, and the German physicist Heinrich Hertz had later shown, that microwaves would behave like light waves. The early radar scientists worked out just how this was so. They had invented a new way of seeing things in the sky. Instead of ray guns, they got radar.

It took an odd character, met with some disdain in the London gentlemen's clubs where the planning went on, to turn the idea into reality.

Robert Alexander Watson-Watt, a pudgy and loquacious man in the Ministry of Defense, pushed the idea of radio detection and ranging. (The British provided the idea but the Americans would provide the acronym.) He tracked a Dutch airliner crossing the Channel in 1937, and by the time of the Battle of Britain he had laid out a network of stations that fed into the underground war room. By the narrowest of margins, and aided by Hitler's and Goering's failure to strike first at airbases rather than at civilians, radar seemed to have won the air war for the British in 1940. "Britain," Watson-Watt declared, triumphantly but prematurely, "is an island once more."

Such security was not long lasting. And America, too, would soon enough no longer be a continent protected from attack by even greater extents of water. With the coming of the atomic bomb, the consequences of bombers crossing the ocean became even more frightening.

Electronics, however, was advancing more rapidly than jet engines or airframes. By the 1960s, the big bomber was an endangered species. By the 1970s, radar had such a lead over even aircraft equipped with their own jamming and spoofing electronics that it seemed unlikely that "the bomber will always get through." This became specifically clear to the U.S. Air Force in the 1973 Mideast war, when some thirty of the topline fighters it had sold to Israel were shot down by improved radar and SAMs. The Pentagon had been right to kill LeMay's B-70—new SAMs would have made it obsolete—but smaller, faster planes were vulnerable as well.

———

In 1975 the Pentagon began convening special conclaves of scientists, engineers, and contractors to consider a response. The Defense Advanced Research Projects Agency (DARPA) was put in charge. Founded in the wake of Sputnik, DARPA served as the Pentagon's version of Bell Labs, a free-thinking outfit dedicated to exploring the frontiers of technology liberated from bureaucracy and interservice rivalries.

It planned the first ICBMs and designed sensors for Robert McNamara's line, the high-tech barrier planned for Vietnam's Demilitarized Zone. It developed the autonomous land vehicle, a huge walking tank, like something out of the Imperial army in *The Empire Strikes Back*. DARPA created improved integrated circuits, sensors, and actuators, the sinews and joints of modern weaponry. But most important, DARPA's funds had built up the computer industry in the 1960s and would give us the computer mouse and the Internet—initially called ARPANET.

To come up with a means of eluding the new, powerful radars, DARPA created a project called Harvey, after Jimmy Stewart's invisible bunny pal, to look into making an airplane invisible to radar, or at least harder to see. It signed up four leading airplane builders and gave them four million dollars apiece to solve the problem. Lockheed was at first not among the four. The irony was that the Skunk Works achievements in reducing radar cross section on the Blackbirds, including the stealthy D-21 drone, had been so secret that no one in the Pentagon knew of them; thus when the discussion turned to stealth, no one thought of

Lockheed. To get the company included in DARPA's stealth studies, along with Boeing, McDonnell Douglas, and Northrop, Rich had to do some fast talking to DARPA's George Heilmeier.

The Skunk Works was not in good odor. Kelly Johnson was seen as arrogant and difficult, living in the days of its past glories. To inform the DARPA scientists of the work the Skunk Works had done nearly a decade before, Rich had to persuade the CIA to release information on the stealthy technology of the A-12 and the D-21. With that information in hand he persuaded DARPA to let Lockheed participate.

———

One day in April 1975, just as he had settled down to a cup of instant decaffeinated coffee, Ben Rich had a visitor. He had taken over as boss of the Skunk Works in January and was looking for projects. Now a young man named Denys Overholser sat down and began to tell Rich about a footnote in a nine-year-old Russian technical paper that had only recently been translated by the Air Force Foreign Technology Division. Bearing the engaging title "Method of Edge Waves in the Physical Theory of Diffraction," the article was the work of Pyotr Ufimtsev, chief scientist of the Moscow Institute of Radio Engineering. It was about radar-evading, "low observable" shapes, what would soon be known as "stealth."[1]

Overholser explained to Rich that Ufimtsev had updated the equations of Hertz and Hermann Helmholtz so that one could for the first time calculate the radar reflection of a two-dimensional shape, such as the surface of an airplane. With that knowledge, you could design an airplane so that it would reflect radar waves away, off into space, instead of back to the receiver. And you could do this regardless of the size of the shape—in other words, a huge shape could be made to look small, almost invisible, on radar.

Overholser, a chunky mountain biker who had shaped radomes for the Skunk Works, was now assigned by Rich to turn the equations into a computer program and the program into the shape of a new airplane.

In five weeks, he and Bill Schroeder, the longtime Skunk Works radar and math whiz who had come out of retirement, wrote a program called Echo to do the calculations of an optimum shape for scattering radar beams. They took the numbers to a junior designer, Dick Sherrer, and by May 5 were back in Rich's office with the results: drawings of an arrowhead-shaped airplane they called Hopeless Diamond.

"So would this one," Rich asked, suggesting for comparison radar signatures given in terms of aircraft or bird types, "be the size of a Cessna or what, a condor, an eagle?"

"Ben," Overholser said. "An eagle's eyeball."

Thereafter, Rich got hold of a number of ball bearings the approximate size of an eagle's eyeball, and took them on his trips to the Pentagon, rolling them across generals' desks and saying, "There's your airplane!"

A few months later, someone on Rich's staff gave him a bowling ball painted TOP SECRET: It was the radar signature of the whole Pentagon after it had been subjected to the Skunk Works stealth treatment.

———

When Johnson saw the sketch of the Hopeless Diamond, he literally kicked Ben Rich in the ass. "It'll never get off the ground," he predicted. Johnson had always said that if an aircraft looked beautiful, it would fly well. It was the classic premise of the great clipper-ship designers and race-car engineers. But this plane was ugly. Rich would write, "No one would dare to claim that the Hopeless Diamond would be a beautiful airplane. As a flying machine it looked alien."

Johnson also loathed electronics, and this was an airplane designed for its electronics, by electricians. "If Kelly could find a hydraulic radio, he would use that," went an old chestnut around the Skunk Works. He was famous for winning his quarter bets on this or that issue of technology; his penny-pinching ways were legendary at the Skunk Works, tokens of his hardscrabble upbringing. Now Rich bet him that the Hopeless Diamond would have a lower radar cross section than the fifteen-year-old D-21. (The calculations suggested it would be a thousand times less visible on radar.) On September 14, 1975, they took the two wooden models of the two aircraft into an electromagnetic chamber. The results were clear, and Johnson handed over the quarter, mumbling, "Don't spend it until you see the thing fly."

In October, they took the model to Gray Butte, the radar cross-section test site that belonged to McDonnell Douglas. On one occasion when the model was on the test pole, there was a sudden blossom of reflection. Uh-oh, the guys in the test center thought, was there some angle they had not considered? Then someone looked out at the model; on its pole in the middle of the concrete, they noticed that a large blackbird had landed on it. Even the droppings from birds could add to the radar reflection—a decibel and a half, as these things were measured, to a total reflection of three decibels.

Then in March 1976 they trucked the black-painted wooden model all the way to the Ratscat—the "radar scatter" facility at White Sands, New Mexico—for a "fly-off" with Northrop's stealth model. The results were so overwhelmingly in favor of the Lockheed model that Northrop radar expert John Cashen was dismayed. The Hopeless Diamond was revolutionary—if it could actually fly. Nor was it clear that Kelly Johnson would be wrong about that. Making this shape fly depended on computers, as the airplane would be too unstable for a mere human pilot alone.

Although Johnson was appalled by it, the Stealth's shape provided the very embodiment of the Skunk Works doctrine of simplicity. "It looked totally alien," Ben Rich had said, because it was radically simple. It was a sculpture on the theme of the cutting away of excess, an airplane that flew no better than it had to so that it could not be seen.

Born inside a computer, it resembled what programmers call a wireframe drawing, making the most of limited processing power. In fact, it resembled the angular tanks and obstacles and flying saucers in the early video game called Battlezone. These shapes would eventually show up in a new kind of aerial combat that itself resembled a computer game.

———

With the machinists union on strike, Skunk Works managers did much of the work on the prototype. The engines were ground-tested at night in a rigged-up barrier composed of two tractor trailers. Then, on December 1, 1977, Bill Park, having demanded and received a twenty-five-thousand-dollar bonus to fly the ugliest airplane he had ever seen—especially after seeing the cockpit, which offered very little space from which to escape—lifted off the runway at Groom in what was now called the XST—experimental stealth testbed—or Have Blue.

Now the Skunk Works had to prove that the real airplane was as stealthy as the wooden model on the pole. Park and other pilots began testing it against real radars, the bad-guy radars, surreptitiously obtained, like the Red Hat squadron's MiGs, and as carefully hidden in the remote corners of the Tonopah Test Range adjoining Dreamland.

During one test, Have Blue showed up as a bright blip on the screen. The engineers couldn't understand what was wrong. Then someone noticed that three screws had not been driven flat. The heads sticking up just a fraction of an inch triggered a huge radar return.

It was soon clear to William Perry, who had become Stealth's champion at the Pentagon, that his scientists were looking at something as groundbreaking for warfare as the jet engine had been, or the machine gun, perhaps even as revolutionary as gunpowder or the crossbow. Keeping this shape hidden was vital. More than any airplane before, perhaps, the form signaled function. You got the idea just by seeing it.

In the late seventies, most of the world thought stealth meant paints or panels that absorbed radar, not the faceting of this airplane. Keeping Have Blue invisible became a top priority.

By day, the strange shape, only three fifths the size of the "real plane," with a cramped cockpit and crude systems, could attract attention. At first the model was disguised in a mottled camou, browns and grays and light blue, but this did not seem to work very well, so it was repainted a light gray. It was never brought outside unless uncleared personnel had been exiled to the windowless mess hall or to quarters, and when Soviet satellites were scheduled to pass overhead it was left in the hangar or under the shelters, called "scoot and hide," beside the taxiway.

————

In June 1977 a small unmarked passenger plane landed on the runway at Groom Lake and from it emerged President Jimmy Carter's national security adviser, Zbigniew Brzezinski. He met Rich and walked around Have Blue sitting in its hangar, and was briefed in a secure room. By the end of the month, Carter had canceled the B-1 bomber and put his faith in the "Advanced Technology Bomber," the B-2 flying wing.

By the autumn of 1978 Have Blue had proved itself well enough that the Pentagon gave Lockheed a contract to build a fighter version. It was to be ready to fly by July 1980. That fighter, despite Kelly Johnson's revulsion, would come to possess a kind of beauty that seemed at first far from conventional standards. It was a model of Dreamland itself: hard, angular, unabsorbing, unforgiving.

In reality, of course, the shape of the Stealth fighter reflected the state of the computer art at the time it was designed (applying Ufimtsev's equations even on a big computer could yield only facet shapes). The calculation of radar reflectance from curves was a more complicated task and showed up in the B-2 bomber and the TR3A Black Manta, or "baby B-2."

By 1980 the word *stealth* had begun to creep into media reports. CBS News television correspondent David Martin filed one pointed piece. It was unclear whether leaks in an election year were a matter of politics, but in August 1980, with candidate Ronald Reagan hitting Jimmy Carter hard on defense issues, Secretary of Defense Harold Brown, with Bill Perry standing beside him, talked about stealth as a breakthrough. Most people then thought in terms of bombers, since the revelation helped cover President Carter's flank, exposed by the cancellation of the B-1. But there was a backlash, too, to the revelations, which the Republicans exploited: Carter should not have revealed such deep secrets.

The formal first flight came in April 1982, but it was another six years before the taxpayer would get a good look at the airplane. With generals and dignitaries lined up along the runway, the airplane began to taxi forward, with Bob Riedenauer at the controls. It slowly lifted off, but only a few feet above the runway it began to veer to one side. The shocked crowd watched as it flipped onto its back and fell onto the lake bed in a huge dust cloud. The rescue trucks rushed up and cut Riedenauer out as he hung upside down. He never flew again.

Mechanics had miswired a new control unit, switching the controls for yaw with those for pitch, so that when Riedenauer had attempted to pull up on the stick he actually sent the plane heeling to the right. Skunkers noted that this was a near duplicate of an accident in which Mele Vojvodich had almost died, in a Blackbird, on the same runway. But the crash was also a reminder that this unnatural shape depended as much on software as on hardware to fly. Its workings were no longer visible, in the manner of mechanical things, but hidden in computer code.

The shape of the craft itself, however, was an immediate signal as to its secret. And that shape would be hidden away in a new secret base north of Dreamland near the town of Tonopah.

———

I drove up to Tonopah one fall day, heading around the nuclear test site and the Nellis range. Las Vegas brags that it is "the city that never sleeps," but I passed acres of new condos west of town, bedroom communities. Can a city that never sleeps dream? Or is its whole waking life a dream, the way the gambler's is—the dream of the long shot?

At the entrance to the Paiute Indian reservation a billboard read CHEAP CIGARETTES. Farther up the road was Indian Springs, the old World War II air base where B-29s took off to drop bombs at the test site, where the Thunderbirds practice—and where Lazar said he was "debriefed."

It looked much as it did twenty or even fifty years before—much as Groom Lake must have looked in the days of the U-2, I thought. But

Warning signs on the perimeter of Area 51, 1993.
Courtesy of the author

Rancher Steve Medlin's Black Mailbox, considered the best spot for saucer seekers outside Dreamland. *Courtesy of the author*

Bob Lazar's tales of working on flying saucers at area S-4 brought tourists to the Little A"Le"Inn bar and restaurant in the nearby town of Rachel. *Courtesy of the author*

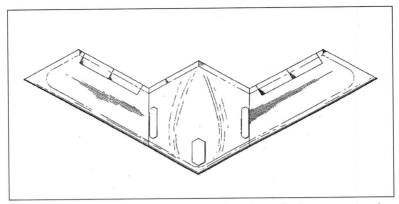

■ This Lockheed patent sketch for a robot aircraft was filed in 1997, several years after "Aurora" was rumored to have been retired. *U.S. Patent Office*

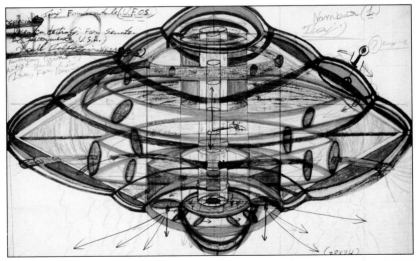

■ *American Dream: UFO Art & Science*, a 1996 flying saucer painting by visionary artist Ionel Talpazan, echoed the sensibility of early saucer "contactees" such as Lockheed employee Orfeo Angelucci. *American Primitive Gallery*

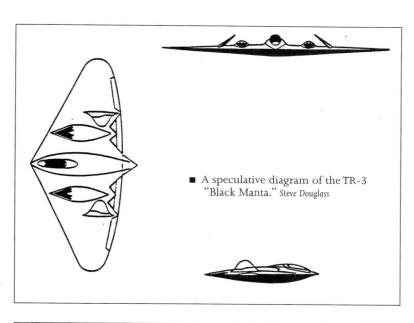

■ A speculative diagram of the TR-3 "Black Manta." *Steve Douglass*

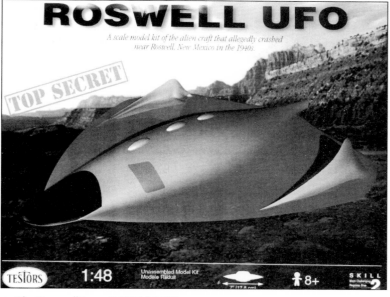

■ The Testors "Roswell UFO," 1997. *Testors Corporation*

■ U-2 spy planes, with fake civilian markings, at "Paradise Ranch" circa 1956; this photo also shows the first crude hangars inside Dreamland. *Lockheed-Martin*

■ The hangar at Roswell Army Air Field where "saucer wreckage" was said to have been stored in 1947. *Courtesy of the author*

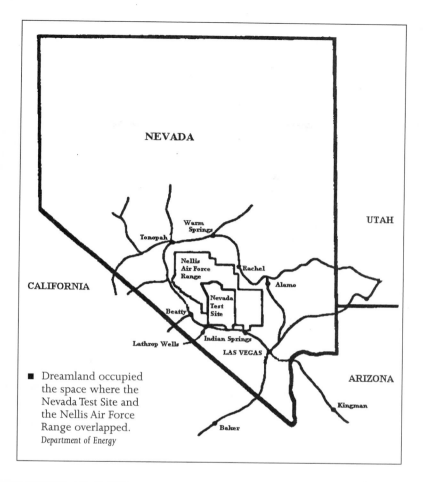

NEVADA

UTAH

Warm Springs

Tonopah

CALIFORNIA

Nellis Air Force Range

Rachel

Alamo

Nevada Test Site

Beatty

Indian Springs

Lathrop Wells

LAS VEGAS

ARIZONA

Kingman

Baker

■ Dreamland occupied
the space where the
Nevada Test Site and
the Nellis Air Force
Range overlapped.
Department of Energy

■ "Red Hat" squadron hangars, used to house captured MiGs inside
Dreamland. *Mark Farmer*

■ The SR-71 spy craft
(ABOVE), later adapted
to launch the D-21
drone (RIGHT), the last
of the Blackbirds.
*Lockheed-Martin; courtesy of
the author*

■ The Darkstar unmanned aerial vehicle at the Lockheed
Skunk Works on the day of rollout, 1995. *Courtesy of the author*

■ "Shamu," or
"Tacit Blue," a
Northrop aircraft
tested at
Dreamland, at the
time of its arrival
at the Air Force
Museum. *Courtesy
of the author*

- A Nevada state highway sign for the "Extraterrestrial Highway," erected near Alamo in the spring of 1996. *Courtesy of the author*

DET 3 SP

JOB KNOWLEDGE

- Cover of the ostensible Groom Lake security manual that the Air Force declared classified in 1996 after its release on the Internet.

- Kathleen Ford's "eyeball," a mysterious shape photographed on Mail Box Road in 1993. *Kathleen Ford*

■ Glenn Campbell's Area 51 Research Center in Rachel, 1996. *Joel Slayton*

■ Steve Medlin's Black Mailbox, reinforced, locked, and painted white, summer 1996. *Courtesy of the author*

there was a recent addition—strange new inflatable buildings beside the flight line—and everything was surprisingly spruced up, as if the base had been restored for a film. (In one version of the Roswell story, Indian Springs had been the site of the first secret saucer storage facility, and perhaps for the storage of alien bodies as well.)

After Indian Springs the four-lane ran out. The sign MERCURY—NO SERVICES introduced the town that had grown from the base camp of the test site. Another sign announced that the fronting stretch of highway had been adopted by NEVADANS FOR PEACE. At first exhilarating, the distance soon became nauseating. I wondered if there could be such a thing as distance sickness, like altitude sickness, the lack of detail and purchase for the eye corresponding to a lack of oxygen.

GATEWAY TO DEATH VALLEY, a sign greeted cheerfully at Beatty. EST. 1903. RIO RANCHO RV PARK. BURRO INN: FULL HOOKUPS. Another pointed to Parumph, twenty-seven miles away, HEART OF THE OLD NEW WEST. And home, I knew, of the Art Bell Dreamland radio show, where conspiracy theorists talked to sleepless callers late at night.

Farther up the road, some of the towns seemed barely worth the trouble of designating them. It was as if the mapmakers had been desperate to work in a couple of dots in the white space so they would not be suspected of slacking off. The signs for the towns told of elevation rather than population, the former being a far more impressive figure.

The mountains now seemed to grow more angular, with pyramidal tops and neatly sliced sides. I had the odd thought that the landscape had crept in as a stylistic influence on the shape of the Stealth fighter, as critic John Ruskin, citing chalets in the Alps, had credited local landscape with influencing architecture.

By the time I got to Tonopah I was in a virtual blizzard. A huge American flag driven by the wind was noisily beating its heart out against a pole in front of the Forest Service office south of town. ELEVATION 6030. HOME OF THE STEALTH.

I stopped at the Forest Service office to get out of the weather and idly looked for maps. A heavy woman behind the counter chatted away. "My husband works out there at the site. Sometimes I go to pick him up, and he warned me that if I ever broke down to just stay in the car. 'Don't get out,' he told me. They don't like people poking around."

At the local historical museum, the exhibits consisted mostly of odd pieces of equipment from the mines, chunks of silver ore, and pieces of crashed airplanes. An aerial photo showed a Stealth fighter flying above the old mining towers and piles of tailings. "Tonopah" means "land of little wood and water" in Paiute, but its mellifluous syllables had become magic to the stealth-chasers.

Tonopah was built on booms, interrupted by busts: the silver boom, then the big booms at the nuclear test site, and finally the sonic booms of secret planes. In 1900, silver was found in the Silver Bow mine, and saloons and casinos and whorehouses were thrown up overnight. In 1922 came the Big Casino, which touted itself as "the Monte Carlo of the desert."

In October 1940 the government turned over some five thousand square miles of public land to the military for training. Government silver certificates replaced the paycheck of the silver miners. A base was in operation at Tonopah by July 1942, and would-be fighter pilots came to the area to learn to fly BT-13 trainers and the P-39 fighter, so dangerous in its handling that both the American Army Air Corps and the RAF had rejected it.

Chuck Yeager was one of the first to train there. He lived in a tar-paper shack heated by an oil stove and recalled that the wind never seemed to stop blowing. "On paydays," he would write, "we crowded around the blackjack tables of the Tonopah Club, drank ourselves blind on fifths of rotgut rye and bourbon, then staggered over to the local cathouse. Miss Taxine, the madam, tried to keep a fresh supply of gals so we wouldn't get bored and become customers of Lucky Strike, a cathouse in Mina, about thirty miles down the road. But we went to Mina anyway, wrecked the place, and the sheriff ran us out of town. The next morning, a P-39 strafed Mina's water tower."

In the fifties, the opening of the test site to the south brought jobs for the miners and other hands. The skills of the miner, by happy coincidence, were in demand at the test site after the Atmospheric Test Ban Treaty was signed in 1962. Testing went underground, and long tunnels had to be built to hold test equipment to record radiation and heat and blast.

Tonopah enjoyed a brief flurry of notoriety in 1957, when Howard Hughes married his longtime companion Jean Peters at the Mizpah Hotel. Hughes picked the Mizpah because he had business in the area to transact. His father had prospected here, and Hughes was betting that the silver veins were not quite tapped out. In a few months, he bought up some 710 claims, covering most of Tonopah along with some 14,200 acres of Nye and areas in other counties, for $10.5 million.

At just about the same time, the nuclear weapons designers at Sandia Labs in Albuquerque cast greedy eyes on the empty areas of the Nellis range south of the old Tonopah base and north of the test site. By 1958, they had set up the Tonopah Test and Training Range, dropping bombs to test fuses and cases and parachutes. The Interceptors would become intrigued by other sectors inside Tonopah's ranges, such as the Tolicha Peak Electronic Warfare facility, Base Camp, to the north of Highway 6 near Warm Springs, and Site IV, where foreign radars are tested, the name an odd shadow of Bob Lazar's mystery site, S-4, at Papoose Lake.

What drew them most, however, was the huge base built almost overnight in the middle of a test range previously dedicated mostly to radar and electronics.

————

One day in 1984, Col. Robert "Burner Bob" Jackson was reading *The Wall Street Journal* when a small advertisement caught his eye. The Chevron petroleum company had secondhand oil patch trailers to sell. Jackson bought the trailers for $10 million; he was about to move the Stealth fighter group from Groom Lake to Tonopah.

The trailers ended up on the raw site south of the old Tonopah Air Field, where mustangs roamed the runways and scorpions crept into buildings. Fences and searchlights went up along the edge of the new base, and video cameras and motion sensors were installed. Eventually the Air Force spent $300 million and fitted the place out with a gym and indoor pool.

The activity did not go unnoticed. A-7s were kept parked outside, as Soviet satellite passes overhead increased to three and four a day. The A-7s were part of a cover story; they carried old napalm canisters painted black and decorated with flashing red lights and lettering that read REACTOR COOLING FILL PORT. The idea was to spread the information that these were an "atomic anti-radar system." Ground crews were forced to lie spread-eagle and not look at the craft as they passed. The lie must be made as hard to get at as the truth.

In the fall of 1988, the Air Force released the first, heavily doctored photograph of the fighter. It was so vague and the angle so misleading that some pilots doubled over in laughter when they saw it.

But soon the airplane buffs found out about Tonopah and the fence. By the winter of 1988, some were getting glimpses, even snapshots, that showed the strange flat shape from the bottom, the angular diamond, faceted and crimped.

Byron Augenbaugh, a schoolteacher and airplane buff from Escondido, California, drove up to Tonopah one day in the spring of 1989. At a gas station he asked where he should go to see the Stealth fighter. "Just look up," the attendant told him, and sure enough one flew over. Augenbaugh snapped a picture. On May 1, 1989, *Aviation Week* ran a cover shot of the fighter so fuzzy that one of the magazine's editors said "it looked like a French Impressionist painting."

There was something lascivious about such images. In the first pictures of the Stealth fighter the Air Force would release, the inlets for the engines were airbrushed out, like the flaws in a *Playboy* centerfold model. Around the same time the Air Force chief of staff went so far as to testify that an airplane, like a beautiful woman, should reveal itself not completely but bit by bit. But artists' impressions of the Stealth fighter and other suspected aircraft that appeared in magazines such as *Popular Science* had their highlights exaggerated, like the women in bomber-nose art, their shapes made fuller and more magical, and with magical light swelling and suffusing the shapes and saturating the colors.[2]

These paintings stood in contrast to the spy photos and flying saucer snapshots, blurry and grainy like the telephoto images of sunbathing celebrities in European magazines—located somewhere between imagination and reality.

The whole experience of snooping for stealth was about the means as well as the ends. It was as much about telephoto lenses, the big binoculars the stealthies called "hooters," or the grainy green mystery images produced by night-vision equipment, as it was about any real craft.

———

Jim Goodall, who had the declared ambition of collecting a picture of every airplane the U.S. Air Force had ever flown, complete with tail number—more than a hundred thousand pictures—claimed an almost sexual rush when he first saw the Stealth fighter in the winter of 1988. For a traveling salesman of computer equipment, he had a surprisingly sensual side, and was a sharp dancer in the Holiday Inn discos he visited on the road. Goodall was often joined on the fence line at Tonopah by John Andrews, the veteran plastic model designer for the Testor corporation. In 1986 Testor had released Andrews's model of the Stealth fighter, called the "F-19" and based on his glimpses as well as reports from the other watchers. It was like putting together a police composite sketch of a wanted man, he said.

The model set off a small storm in Washington. How could a model company know what America's most closely guarded secret looked like, when our lawmakers themselves did not know? Of course, everyone in and around the black world knew that the last person to be briefed on a project of such high security was a congressman. You might as well just publish the specs in the *Congressional Register*. Angered and embarrassed, congressmen held hearings to find out how the shape of the plane had leaked. By one account, the Air Force had to bring a model of the real fighter to Capitol Hill in a locked box, handcuffed to a guard, to illustrate to them that Andrews's model was wrong.

But the Air Force and the Skunk Works could only say it was wrong, and not show it, unless they broke down the very secrecy designed to keep people like Andrews out.

————

I drove east out of town toward the base. The road was lined with corrals and horse stables indistinguishable from houses, and old mines and piles of tailings. I passed a truck with a bumper sticker that advised IF IT DOESN'T GROW, IT HAS TO BE MINED.

I was looking for the base, which did not appear on the maps. The old World War II main base is the civilian airport now. On the official Nevada state highway map I had picked up at the Forest Service, the whole area was vaguely named "game range." Even the test site and the Nellis range were omitted.

As the buildings thinned out and then stopped, the sight line shrank to a few hundred yards. The road was marked "Grand Army of the Republic Highway." "Ely 163 miles," I read, "next gas 112 miles." Soon I saw it to the south: the old World War II base, now Tonopah Airport, with huge arched hangars and earthen bunkers suffused in the soft light seeping through the black clouds.

I drove along the near deserted flight line. It seemed dark, almost haunted. It had been from the beginning a kind of hard-luck base. Trainees, suffering from the cold, the wind, and the dust, named it "Camp Frosty Balls" and died at alarming rates. After the base became the site of a B-24 training program, the bombers kept crashing, too. Once, a machine gun began firing inside one of the bombers, and when the crew finally got it on the ground, two men were dead in the turrets.

By the end of the war they were testing bizarre bat bombs, crude radio-directed cruise missiles that foreshadowed the future of the base.

———

By the time the Stealth arrived in 1984, the Tonopahans had learned the importance of noting the side upon which their bread was buttered. Their conspiracy of silence about the secret airplane was like that of a beach town at the height of the season when a shark is sighted. They paid no note to the airplanes flying overhead.

Taking off at night, Stealth pilots, like SAC pilots before them, practice-bombed America in the dark, targeting boat docks in Minnesota and high-rises in Denver. "We could find Mrs. Smith's rooming house and take out the northeast corner guest room above the garage," one of the pilots boasted.

Still, even those who had trained with the plane were not sure it would work against real radar. Before the Gulf War, the general sentiment among American pilots in Saudi Arabia was "I sure hope this Stealth shit works." Then they saw the bats that showed up each morning dead on the floor of the hangars, their sonar fooled by the faceted shapes of the planes, just as radar would be, and they believed.

On the night of January 17, 1991, a retirement banquet was held to honor Ben Rich. Halfway around the world, the F-117s were loading up to hit Baghdad.

Soon they would be going after another target—the press. Peter Arnett of CNN, whose coverage was unbeloved by the Air Force, was using Baghdad's phone center. The switchboard went on the target list, and one night, in the ready rooms at King Khalid Air Force Base, off-duty pilots waited expectantly, sets tuned to CNN. They counted down the seconds until, right on schedule, their screens suddenly went to a roaring gray and cheers broke out.

By the time the war was over, the F-117 was a national hero, and the pictures were no longer distant and grainy. The Stealth was photographed by Annie Leibovitz, photographer of the stars, for *Vanity Fair*, in a portfolio together with Schwarzkopf and Powell and Cheney—war celebrities. Seeing the airplane there reminded me of those portraits of American Indian chiefs, hauled across the Atlantic to entertain the court in London, so strangely out of place. Stealth was now seen openly and it possessed the beauty of the jeep or humvee, or the pup tent or camouflage.

The sort of claims the airpower advocates always like to make were now made for the F-117, that in just so many sorties the Stealth had done the work of the entire bomber fleet of World War II. It reminded me of Curt LeMay, bragging about the B-47 or the B-58.

After the F-117 was made public, the locals could show their pride openly, and after the Gulf War, Tonopahans held a victory parade with a thousand people. I drove past the fire station, which bore a bas-relief of the fighter and a "Home of the Stealth" plaque.

Yet as soon as the Stealth became a hero, it was gone. The whole wing of aircraft was transferred in 1992 to Holloman Air Force Base near Alamogordo, New Mexico. The base at Tonopah was too distant, and keeping the planes secret was too expensive.

———

Back on the highway, I was soon spun into a cocoon of snow. As if by meteorological conspiracy, the whole place had locked down. I drove up to the fence, paused a reflective moment or two, then turned around.

Back in town I stopped for coffee and cherry pie—$1.50 total—a few feet from the fire station. At the pay phone, I noticed a plaque on the wall that showed the outline of the state of Nevada with Iraq laid over it. In the center, roughly where Dreamland stood, Baghdad was neatly superimposed and marked by red flashes. "First to strike, January 1991," the plaque stated proudly. I was confused about the scale: Was Iraq that small, or Nevada that large?

———

On the fence line at Tonopah, the guards were usually polite and friendly enough, unlike the camou dudes at Freedom Ridge, but later the Interceptors discovered another viewpoint, which for obscure reasons they named Brainwash Butte. From there, you could see the base, but the view was not a very exciting one: The long row of identical hangars that had been built for the Stealth fighters looked from this distance like the little tin sheds of one of those U-Store-It rental facilities.

Goodall was convinced something new was going on. The security was tighter than it had been during the height of the Stealth program, and new construction was under way. Goodall should have known. He had been venturing to the perimeter, both at Tonopah and Groom Lake, longer than nearly anyone—anyone except for a very strange man named John Lear.

13. The Decentral Intelligence Agency; or, "Use of Deadly Farce Authorized"

John Lear's telephone answering machine does not give his name or number. But in his voice it offers the following: "To leave a message for Area 51, push one. To leave one for S-4, push two. The Tonopah Test Range is temporarily unavailable."

Lear comes on the line, fumbling. He has flown more than 160 types of aircraft in fifty countries, but he can't figure out this damn machine and he jokes about it. Most often these days, Lear can be found in the Holiday Inns surrounding distant airports. A commercial pilot, he has had a hard time keeping a job since he became one of the most visible viewers of Dreamland and proponent of UFO theories.

John Lear journeyed to Dreamland in a Detroit dream machine. In September 1978, he got behind the wheel of his Lincoln Mark IV and drove to the edge of Groom Lake. At that time, the perimeter still ran along the lake edge, and the mountains and road were still public land. Lear had long known about the base, about the U-2 and the Blackbird, and now he had heard rumors that something else was flying. Ahead of him, the rank of hangars that once held the Blackbirds were visible along with a few aircraft—a MiG on the flight line, a transport.

He quickly snapped off a few photos and waited. "Then a half hour later this Klaxon goes off and we see a little trail of dust." Two vehicles, heading his way. "I rolled the film up and put it in the ashtray of the Mark IV and put another roll in the camera and shot the same thing again. A black guy in a red car came up shouting, 'What in the hell do you think you are doing?' I decided to play it cool as possible. 'So we're not supposed to be here, right?' I said.

"Then I went into a whole line of BS. My dad did the autopilot for the U-2, and I've got a lot of good friends in the SR-71, and so on. I used to live near the airport in Burbank, and we would always see those three Constellations that went up here."

The guard calmed down. "Do you have film?" he asked. Lear pulled the film out of the camera and gave it to him.

He promised not to intrude again and was allowed to leave. He promptly drove to Los Angeles, had his stashed film developed, and made big 18-by-20-inch prints. His black-and-white panorama of the base from across the dry lake, then covered with a thin layer of water, would become famous, although so many buildings have been added at Groom that today the picture makes the place seem crude and primitive.

———

"Truth," Lear once wrote. "I can't tell you what the truth is . . . I'm not sure such a thing exists. If it does exist, the truth is hidden in an incredibly complex, labyrinthine hall of mirrors with floors of quicksand leading to truly frightening bizarre and awesome events which have been going on for billions of years, if not eternity." John Lear's writings are apocalyptic, almost hysterical, but in person or on the phone he is charming and reasonable.

John Lear is the son of Bill Lear, the aviation pioneer who created some 150 major innovations in radio and control systems, along with the eight-track audiotape and the jet that bears his name.

Born in 1942, before war work made his father rich, John was alternately spoiled and abused. From the age of twelve he could barely bring himself to speak to his father, and family meals terrified him. His father would begin by speaking tenderly but quickly rise to a harangue over some failure of John's. Once Bill Lear, dismayed with John's ducktail haircut, slapped him.

The Lears spent a lot of time in Switzerland. Bill Lear nicknamed their estate there Le Ranch. John was rarely in any school for more than a year and was eventually sent to Le Rosey, the posh Swiss academy known as "the school of kings."

John Lear was obsessed with flying—perhaps because his father, for all he had contributed to aviation, held the lowest possible regard for those who actually flew planes for a living. He made his first flight at fourteen, in 1956, and got his license and soloed at sixteen. He immediately declared his intention to become a commercial pilot. He added twin engine, instrument, and aerobatics ratings. In December 1960 his father's company, Lear International, hired him as a public relations representative and pilot.

Then, on June 24, 1961, to get to Bern, Switzerland, on an errand, John rented a small yellow single-engine biplane from a flying club in Geneva. He had often made low wing-wagging passes over the dorm at Le Rosey in his Cessna, and now he came across again, ready to put on a show of aerobatics.

Screaming like a rodeo cowboy to the students below, he began a three-turn spin at well under a thousand feet, intending to pull up just a few feet from the ground. After the second turn, with his nose pointed to the ground, he realized he was too low. He saw a barn out of the corner of his vision. He began to pull back on the stick, but the plane was still heading down at a 30-degree angle when it plowed into a wheat field, smashing him into the instrument panel and snapping the straps of his shoulder harness.

Students pulled him from the wreck. In the ambulance, doctors performed an emergency tracheotomy. Lear's larynx had been crushed. Both sides of his jaw were broken, four front teeth were gone, his heel bones and ankles were crushed, and each leg had been broken in three places. He spent five hours in surgery in a Geneva hospital and several days in intensive care. His father, angry and humiliated, came to the hospital immediately, but never returned during Lear's long convalescence.

In 1962, Lear agreed to attend Art Center College in Pasadena, but lost the $5,000 his father had given him for tuition on a stock tip. In 1964 he was part of a crew taking a Learjet on a round-the-world flight. The crew went east, violating Indian airspace, making their longest leg— into Singapore—with only enough fuel for three minutes in the air. A MiG-17 shadowed them near the Kuril Islands, then flew off when one of the crew raised a camera with a telephoto lens and began shooting through the Plexiglas window.

Bill Lear warmed to his son after that flight, but the breach was never really healed. His will was generous to John's children—they got 15 percent of his fortune—but John was left out. At the funeral, John Lear cried uncontrollably. He eventually became a pilot for Air America—the CIA's clandestine airline in Southeast Asia—as well as for domestic carriers.

Growing up in California, Lear was aware of black programs; his father's company supported some of them. He knew about planes that flew workers and equipment from Lockheed in Burbank to the Ranch. In the mid-seventies, Lear heard rumors from a reporter friend that more interesting things were going on at Groom Lake. In those days there was practically no security, and Lear was able to drive almost to the lake itself. "That's when I took that famous picture of the lake bed."

After 1978, Lear became increasingly fascinated with UFOs. He would eventually drop out of MUFON because the organization wasn't hard-core enough for him. He grew close to those who searched for black aircraft, but also to the UFO believers. In 1987 he published his "Darkside" thesis, the most extreme view of the dark dangers of aliens, full of tales of secret treaties with the aliens and their need for human and cattle bodies.

Lear came to believe it all—the underground bases, the tanks with aliens and alien-human hybrids, the bases on the moon and Mars, MJ-12 and the secret treaties. He even went on record as believing George Adamski, the early contactee.

He claimed that aliens mutilated cattle to extract a special enzyme. "The secretions obtained are then mixed with hydrogen peroxide and applied on the skin by spreading or dipping parts of their bodies in the solution. The body absorbs the solution, then excretes the waste back through the skin. The cattle mutilations . . . were for the collection of these tissues by the aliens."

At the Ultimate UFO Conference in Rachel in 1993, Lear declared, "In 1979, our alliance with the aliens became a disaster . . . Forty-four U.S. scientists and approximately sixty-six members of Delta Force security personnel were killed by the aliens in an altercation at a jointly occupied U.S.-alien base north of Los Alamos, New Mexico . . . The exact cause of the altercation is not known, but the cause of death was listed as external head wounds. This effectively terminated the alien alliance for an indefinite time."

The gray aliens of the Lore were simply robots working for a race of aliens that resembled praying mantises. The government had tried to prepare the public for release of information on the secret treaties by sponsoring such films as *E.T.* and *Close Encounters of the Third Kind*, but then relations went sour. MJ-12 was in disarray and confusion. It was time for the truth to come out, Lear cried.

———

Lear's transformation struck many among the Interceptors as suspicious. He had worked for the CIA in Southeast Asia. He knew many of the Interceptors. He had introduced Bob Lazar to the newsman George Knapp, who publicized his story. Lear kept popping up in the stories about Area 51, smack between the stealthies and the youfers, working each way. It was easy to see a scenario where he was a disinformationist. He had, after all, worked for Air America. But Lear loved to fly; he would fly for anyone.

Yet another explanation seemed more convincing. "He has no bull-shit filter," one of the Interceptors has said. He was both totally credulous and totally suspicious: Lear never met a plot he didn't like.

Today Lear no longer wants to get the story out. He doesn't think the public is ready. Those who are keeping it all secret know what they are doing. John Lear, who once challenged the government to come clean, now thinks they may be right. Yes, he says, there may be disinformation. There may be government influence in the media. Look at all the films on UFOs. He doesn't even trust supermarket tabloids. Somehow they get information early. They manage to take the kernel of truth and distort it just enough to make it look ridiculous.

Was that, I wondered, the case with a story like the one headlined, TOP SECRET: U.S. HOLDING NAZI WAR CRIMINALS IN SECRET AREA 51 IN NEVADA—AS SLAVE SCIENTISTS TO BUILD WEAPONS! I went away shaking my head. I had never thought to suspect that the government might control the *Weekly World News*.

———

The first time they climbed Whitesides Mountain to survey Area 51, Jim Goodall thought Lear would never make it because he has flat feet. But Lear carried the sixty-pound pack all the way, and Goodall was the one who had the hard time.

Goodall may have been the most fervent of the Interceptors. As I had, he had grown up in the shadow of the SAC B-36.

One evening in the summer of 1951, when he was five years old, Goodall felt his father shaking him awake. There's something you've got to see, his father told him. Young Jim went outside and heard the rumble of two dozen B-36s and saw their shadows—"aluminum overcast." He was fascinated. When the family moved to the San Francisco area, he found his way to airplanes again. He once sat in the prototype XF-104 Starfighter in a wind tunnel in Sunnyvale and managed to close the canopy. Even as a kid he knew enough to be careful of what lever he pulled; he knew there was such a thing as an ejection seat. Another characteristic of his personality was already forming: He talked his way out of trouble.

He joined the Air Force in March 1962, and in February 1964 he was at Edwards working on a communications system. President Johnson had just made the existence of the Blackbird public, and Goodall saw his first, a YF-12. He still remembers the date—February 29, 1964. "I was about to get on the Northrop shuttle to Hawthorne when I heard this incredible roar and ran down to the flight-line area and looked to the south."

There Goodall saw a black airplane that he at first thought was the famous X-15 rocket plane, but from the scale of the people standing beside it he realized it was a larger craft. The little prop shuttle took off and it flew right over the taxiing YF-12. The moment he saw the Blackbird framed in the window beneath him, he realizes now, he imprinted on it like some infant animal. He was locked in to the fascination of his life. "I could not believe my eyes," he remembered later. "At that point I became obsessed."

After he got out of the Air Force, he would split time between selling computer hardware on the road and serving in the Air National Guard in Minnesota. As unit historian, he managed to persuade the Air Force to provide him with an old Blackbird for the group's museum. It was an A-12, an agency plane, and Goodall made it the most meticulously restored and maintained Blackbird in the world. In time, Goodall was admitted to the Roadrunner's Club, whose members had worked on the U-2 or the Blackbirds between 1955 and 1968 at the Ranch.

He has calculated that he's spent some eighty days on the perimeter—twice the time Jesus spent in the desert—on Whitesides and Freedom Ridge, then by the fence line at Tonopah, looking for the Stealth fighter, and later at Brainwash Butte. He would take one of the first clear pictures of the F-117.

By the time he went up to Whitesides to look down on Dreamland for the first time with John Lear in the fall of 1988, his obsession had expanded. At some point during the revelation of the Lazar story, and talking to those who had worked at the base, Goodall crossed the Ridge—or began to straddle it. He came to believe in the presence of alien craft, as did John Andrews, his frequent companion on the trips. "There are things out there that would make George Lucas green with envy," he had been told, and he believed.

The key moment in his conversion was a letter Ben Rich had written to him, in which Rich said that both he and Kelly Johnson believed in UFOs. (But in the account I had, this was a tease.) Goodall talked often with Rich, who respected him as a true buff, someone who saw that what the Skunk Works had done was important history. Rich even appreciated the efforts of Goodall and the others to get the story out; as he grew older, he saw that the whole system of secrecy had grown more and more onerous. Rich now felt that it was out of hand, and he once compared the Interceptors to Ross Perot, shrilly crying for a change in a system gone wrong.

Goodall had come to believe in the saucers. Something, he wasn't sure what, had happened at Roswell. He could believe most of Lazar's story. Perhaps Lear—as always, a central figure, the key link—had influenced him, but what for most of the Interceptors was just an intriguing possibility became a certainty for Goodall.

It did not reduce his interest in black craft. He was still into every detail of every possible project. He became the butt of gentle jokes about his constant obsession with "something new at Tonopah." He would hide under camou net for days and come back reporting that security was tighter than during the Stealth deployment and that some new craft must be flying. But he was not able to pin down what craft.

John Andrews was constantly enraging the people at the Skunk Works. The very mention of his name, and his constant letters of inquiry, sent Ben Rich fairly raving. Kelly Johnson had been outraged when he learned in the early eighties that Andrews had been allowed to photograph and measure the D-21 Blackbird drones in storage at the boneyard at Davis-Monthan—the same strange shape I was told I did not see.

In 1959 he knew all about the U-2 and contacted Lockheed, but he honored the company's request not to produce a model. Only in 1962 was a miniature U-2 released by Hawk Models in Chicago.

When Andrews was pursuing the Stealth fighter, an AFOSI officer flew out from Washington and told him, "Just be patient." Andrews expects AFOSI to keep an eye on him; it's their job. But today, Andrews feels, "things have changed. Once it was man to man. Now they are hiding behind regulations."

When his model of the Stealth fighter, billed as the "F-19," appeared in 1986, it became the best-selling plastic aircraft model of all time, with a million sold, and it is now highly sought by collectors. Although Andrews estimates its dimensions were accurate to about 2 percent of the real thing, and 75 percent accurate in shape, in fact it turned out to more closely resemble the speculative Russian Stealth fighter, the experimental MiG Ferret. But some of the buffs, who had long imagined the craft, would later say it looked more like the *idea* of Stealth than the real one.

Andrews was unapologetic. "The model helped keep the security of the airplane, because everybody was looking at it, saying, That's what it looks like."

"But," I interposed, "what if you had been more accurate?"

He had no answer.

———

Andrews next turned out his model of the long-rumored "Aurora" spy plane, with its pulser engine. This came directly from his visits to the perimeter. "I've slept on the top of Whitesides," he said, "and heard the pulser in December 1992. You cannot mistake it. It has a very low frequency; there's nothing like it."

To some of the Interceptors, though, the appearance of the Testor company's Lazar saucer showed that Andrews had crossed the line.

"I'm quite comfortable with Lazar," Andrews has said, and he seems to believe most of Lazar's story. He'd consulted with Lazar and Jon Farhat, a computer graphic designer who was working on the long-gestating film about Lazar, in the development of the model.

Andrews's model of the Lazar saucer was skillfully packaged so that no one could tell just how seriously it was intended. "Area S4 UFO Revealed!" ran the copy on the box. "A scale model kit of the alien craft allegedly hidden in Nevada by the U.S. Government as described by eyewitness and former government physicist, Bob Lazar." Paint and cement not included. Skill Level Two. Sixteen-page full-color book included. "Type of vehicle: Anti-matter reaction, gravity amplification, interstellar craft. Made of metallic substance of unknown nature, containing an antimatter reactor to bend space-time, fueled by element 115." Rendered in 1/48 scale, it was made up of twenty-three plastic pieces, including a transparent top to offer a view of the antimatter reactor. Testor also carefully stated on the box that "we can neither confirm nor deny" the existence of the craft on which the model is based.

It was the saucer Lazar had nicknamed "the sport model," and it sold out immediately, thanks perhaps to the fact that Larry King displayed the model on his desk during the October 1994 program he filmed from outside Area 51.

The Testor model made Lazar's tale tangible. Once one had seen such detailed plastic parts, it was harder not to believe in the existence of the real craft. Andrews seemed to buy into the "trickle out" theory—all those bits and pieces, they were what the government wanted us to know, so we would be less shocked when the whole truth comes out.

As Andrews's interest in flying saucers grew, his letters to Ben Rich and others at the Skunk Works irritated them even more. Then Rich finally sent Andrews and Goodall that letter in which he admitted, "Yes, I believe in UFOs, and so did Kelly Johnson.

"Yes," Rich continued, "I call them *UnFunded Opportunities*"—in other words, Lockheed ideas the damn fool Air Force wouldn't pay for. It was a joke, and not a kind one.

————

After he finished the Stealth fighter model, Andrews began to hike up Whitesides Mountain, sometimes with Lear and Goodall. Now he was looking for Aurora, or whatever it was that left the doughnut-on-a-rope contrails. After PsychoSpy moved to Rachel and began to publicize the viewpoints, the numbers of viewers grew. As in complexity theory, the first individuals evolved into a self-organizing group of watchers who would later call themselves the Dreamland Interceptors. The name was taken from the *Intercepts* newsletter Steve Douglass published for the secret-aircraft buffs and military monitors who eavesdropped on aircraft radios on their scanners.

Andrews, having watched black planes since the days of the U-2 and having been out on the perimeter since 1988, came to be viewed as the most veteran and venerable of the Interceptors. "It's like a little CIA out there," he said. "We collect bits and pieces and put them together in a mosaic."

The Interceptors developed their own loose camaraderie and culture over the course of many visits. As their totem, the Interceptors adopted the aluminum lawn chair—that icon of suburban backyard America. It was one thing to say you had seen the base—everyone somehow seemed to feel, doing it, that they were among the first, the proud, the few—but the real badge of honor was to carry that chair up there.

USE OF DEADLY FORCE AUTHORIZED, read the signs on the perimeter, citing the Internal Security Act of 1950—also known notoriously as the McCarran Act, named, as is the airport in Las Vegas, for Nevada senator Pat McCarran, although it was promoted and written mostly by then Congressman Richard Nixon and Senator Karl Mundt. It struck me as appropriate to think of Richard Nixon writing the perimeter warnings.

The law's language includes one of the clearest and most specific statements of the outlook and assumptions of the Cold Warrior:

> There exists a world Communist movement which in its origins, its development, and its present practice, is a world-wide revolutionary movement whose purpose it is, by treachery, deceit, infiltration into other groups (governmental and otherwise), espionage, sabotage, terrorism, and any other means deemed necessary, to establish a Communist totalitarian dictatorship in the countries throughout the world through the medium of a world-wide Communist organization.

Before Glenn Campbell discovered Freedom Ridge in 1993, the best view of Dreamland was from Whitesides Mountain; farther away, after Freedom Ridge fell victim to the expansion of the perimeter, there was Tikaboo. Tikaboo became the agreed-upon standard for the measurement of the height of other peaks, the strenuousness of other hikes, and in planning expeditions to observe the base at Tonopah, the nuclear test site, miscellaneous mysterious electronic stations, and sites of aircraft wreckage. Visitors speculated on areas they could not reach, such as the fabled Cheshire airstrip, which was said to remain invisible until special lights were turned on. Or Base Camp, a mysterious facility north of Warm Springs and Highway 6, or Site IV, deep in between Tonopah and the restricted area around Groom. It was the home, Agent X reported, "of terrain-following radar development, covert testing of purloined Soviet, Warsaw Pact, and Chinese radars and ECM . . . making sure that they wouldn't jam fuses on our nuclear weapons and disable our penetrating bombers' electronic navigation and countermeasures. It seems to be an

integral part of the Nellis Range Complex electronic warfare and evaluation capabilities along with the Tolicha Peak Electronic Combat Range."

The mock spies—"a little CIA"—and the jesters of Dreamland
watched the reliquary of the Cold War with whimsy and cynicism. They
wore the same camou as the camou dudes. They reminded me of Marx's
famous statement that history happens twice, the first time as tragedy, the
second as farce. Their production was a send-up of the Cold War; their
spirit like that suggested on the old Firesign Theatre album cover bearing the revolutionary banners of Marx (Groucho) and Lennon (John).

Of course they were also just another of those self-directing
American groups Tocqueville had observed, revealing a nation of joiners
and near-obsessives. I recognized myself in them: We had been the kids
who put together too many aircraft models and spent our time at the library looking at *Aviation Week* instead of reading the Hardy Boys.

Among them were journalists and buffs, private researchers and
conservationists. Peter Merlin, an aviation archaeologist, found the crash
sites of old planes in the desert and ferreted out details and documents.
He carried a key ring made of bits of famous planes he'd found. Tom
Mahood, the former civil engineer from Irvine, spent days assembling
careful chronologies and descriptions of secret places like the radar
cross-section facilities. He collated official brochures about Tonopah and
studied old maps.

Agent X, a former Coast Guard agent and reporter for such magazines as *Gung Ho* and *The Nose*, came from his home in Juneau, Alaska, and
spent days driving around the perimeters in rented convertibles turned
into makeshift off-road vehicles. The cars usually came back to the rental
agency in appalling shape. Once Agent X wrecked a Buick LeSabre on a
cutoff from Groom Road, sliding into a ditch doing 60 miles per hour.
His report made it sound like a crash of some exotic secret prototype:
"The LeSabre rose in a 45-degree left roll before hanging for a moment
and falling back to the desert floor."

The Interceptors had no clubhouse or Raccoon Lodge. Their social
organization could be described as ad hoc. They had no regular or official membership and only the most general of shared values and beliefs.
They were against excess secrecy, but without the mystery they wouldn't
have been on the perimeter.

Some derided the youfers, some were curious and tentative, and for
many the saucers stories were a little pilot flame of possibility that kept

them all going—the Biggest Story in the History of Mankind. "I'm a hardware guy," Jim Goodall said, but he was willing to speculate on the existence of extraterrestrial hardware.

Many of the Interceptors admitted that people with more vital social and personal lives did not end up hanging around the perimeter. Some saw looking for airplanes or secret saucer bases as just another way to get outdoors, camp, get some fresh air.

There were those who would fly in light planes around the perimeter, an enterprise that felt daring and exciting but offered very little new perspective or information. They visited places like Mount Charleston, where the wreckage of the C-54 that crashed in 1955 on its way to Groom Lake still lay tangled. The circle of the Interceptors widened. In August 1994, some sixty people mustered on the Ridge for what was billed as Groomstock, which included a former pilot from the Blackbird program and UFO buffs.

Once, some of the Interceptors arranged to take a tour of the Nevada Test Site with Derek as their guide. They were taken to the Command Post, with those Naugahyde chairs, the long oval tables, the maps and video screens, and given box lunches. Derek seemed in a hurry to get them into the post, although there was nothing in particular happening, no special event. Only later did it strike Bill Sweetman that they had been carefully kept inside so that they would not see something flying overhead. When he asked Derek, "What kind of pumpkins would we have turned into if we had been outside at noon?" Derek was sheepish.

After the tour, they drove all the way down to Los Angeles and crashed at the Minister's place in the Hollywood Hills in the wee hours of the morning—only to be tossed awake a couple of hours later by the big L.A. earthquake.

The demographics of the Interceptors tended to overlap those of engineers, as the test site overlapped Dreamland. The techie connection brought with it a cynicism that resembled the mercurial charge of a semiconductor. It faded in and out, suspended possibility and speculation like a force field.

The group pursued knowledge by the accumulative and comparative methods of any good intelligence agency. For the Interceptors, simply laying out the known and marking where the unknown began— patrolling the perimeter, so to speak—was enough. Some compiled elaborate grids and tables recording information about sightings, crashes, types of aircraft known and suspected, sometimes right down to tail numbers. In this they had much in common with obsessives anywhere on the planet.

If the Interceptors parodied the CIA's assemblage of information from bits and pieces, they also parodied its surrogate identities and the cult and camaraderie of secret military units. They took on alter egos, in the manner of blues musicians or gangsters, with names to match, mostly e-mail nicknames. The alter egos both resembled and mocked cover names of the UFO informants "Condor" or "Falcon." They created insignia—pins and patches—as if they were a real military unit. They developed their own personal subset of the Lore.

"It's about two and a half Tikaboos," an Interceptor would say. "But how far is it LeBaronable?" the response would come, the rented Chrysler convertible favored by Agent X lending a mobility standard for the dirt roads in the area.

One of PsychoSpy's sources declared that the aliens had bequeathed technology to humans through Hungarians—atomic physicists Leo Szilard and Edward Teller, aerodynamicist Theodore von Kármán, weren't they all Hungarian? And the aliens spoke a language like High Hungarian. This was a twist on an old joke from the Manhattan Project, when someone noted how many Hungarians stood among the top ranks of the project scientists—and how strange their Magyar tongue was. They must be from Mars! The joke delighted the Interceptors, and they enjoyed adding details to it. Visiting Budapest, Glenn Campbell found more than four hundred Lazars listed in the phone book.

Agent Zero had special decoder rings machined up that purported to translate Hungarian into English. They were made of titanium, just as the Blackbirds were, then anodized in a special secret fluid to lend them a silver-blue sheen that seemed to me to reflect the whole happy glitzy fascination with Area 51. The special secret fluid was Coca-Cola.

It was like putting together a mosaic, John Andrews had said, and mosaic was just what the military feared, how they justified concealing the smallest detail. Mosaic was right in another way: Mosaic was the name of the first browser program for the World Wide Web. The information Interceptors gathered found its most natural home on the Internet, in alt.conspiracy.area51, the Skunk Works digest, and, later, on elaborate Web pages.

It turned a lot of watchers into philosophers. A Washington State journalist named Terry Hansen published his musings on the Internet under the title "The Philosophy of Dreamland," taking an epigraph from Jim Morrison ("On the perimeter, there are no stars"). "So set aside your heartfelt prejudices and incredulity for the moment," Hansen intoned, "and come along on an epistemological adventure into the tangled and shadowy jungle of officially forbidden knowledge. Here, rational analysis can no longer be considered a reliable guide. This is a realm ruled by the high priests of the intelligence community who simply do not like us poking our noses into their business, even though we're footing the bill for it . . . Any hopes for certainty must be left behind at the outer boundaries of consensus reality, for we are about to explore the enigma of Dreamland."

As high-technology shrine or secret saucer base, Area 51 took on an existence on-line as a virtual place, a "notional" country. The buffs were rebuilding Dreamland on-line in HTML and reverse-engineering it in data.

It was Vannevar Bush, the very man reputed to have been head of the secret Majestic, or MJ-12, group, who first laid out the vision of the personal computer and its new ways of organizing information. In July 1945, in a celebrated article in the *Atlantic* called "As We May Think," Bush described his vision of the Memex, a personal memory or information device. In the process, he projected something that sounds like CD-ROM and the Internet. Bush's ideas would inspire those who created the personal computer and the Internet, people like Douglas Engelbart, inventor of the computer mouse, who read the *Atlantic* article in a straw hut in the South Pacific when he was in the military. Bush believed that the human mind operated less by classification and organization, the traditional view of thought, than by association. "With one item in its grasp, it snaps instantly to the next that is suggested by the association of thought, in accordance with some intricate web of trails carried by the cells of the brain. Yet the speed of action, the intricacy of the trails, the detail of mental pictures, is awe-inspiring beyond all else in nature."

Bush had predicted, "Wholly new forms of encyclopedias will appear, ready-made with a mesh of associative trails running through them, ready to be dropped into the Memex and there amplified." He predicted the rise of a special profession of innovator to mark such trails and distribute them to individual Memex machines, "a new profession of trailblazers, those who find delight in the task of establishing useful trails through the enormous mass of the common record." This anticipated the strategy of the Interceptors. They were trailblazers to documents and information of an unusual or speculative nature.

The assemblage of extremely detailed and factual studies, notes the historian Richard Hofstadter, looking back over American history in the light of McCarthyism, has long been characteristic of conspiracist groups. From these sometimes rickety, jerry-built edifices of fact, the great leap to a wider conclusion is made—a leap of suspicion that is the dark side of a leap of faith. The information the Interceptors would assemble in time lines and lists and collections was like nothing so much as a conspiracist assemblage. But not everyone was willing to climb to the top and make the leap they had prepared.

———

The Interceptors knew something was happening when they showed up in the movies. The 1996 film *Broken Arrow* contained a reference to "the guys in lawn chairs just watching for something to take off." In the film, a bomber and a nuke go astray, and the first official instinct is to cover it all up, keep it secret. But, a young aide warns, "Don't forget the guys in lawn chairs" watching all the bases, who will notice a bomber leave and not return. Area 51 had crept into the mainstream of popular culture with a speed that surprised even the hardiest Interceptors.

———

The de facto leader of the Interceptors' Nevada branch, Glenn Campbell, aka PsychoSpy, aka the Desert Rat, aka "not the singer," was a controversial figure in the town of Rachel. He irked some locals by failing to commit clearly on the UFO question, and some Interceptors were dismayed by his limited interest in secret aircraft. The saucers were taken as certainty by the Travises and Chuck Clark. It was popular among the Interceptors to see Clark as Campbell's opposite, and to view PsychoSpy's Research Center and the Inn where the Travises and Clark held forth as two poles, one honest and inquiring, the other opportunistic and mercenary. But it was not so simple: They seemed to be distorted reflections of each other rather than opposites. Campbell's patch and guide were also commercial enterprises, and Clark's efforts were certainly philosophical. Campbell needed the black aircraft that bored him; Clark and the Travises, playing out their interview roles over and over, had come to need the saucers. One sometime resident compared them to the two drug connections rumored to be found at either end of town, competing with each other in the sale of crystal meth, a popular antidote to desert ennui.

But if he figured as the Hamlet of the hamlet of Rachel, Campbell was also the village explainer, laying out the mythologies, systematizing the lore. Glenn Campbell was the closest thing in Rachel to Joseph Campbell. In his role as PsychoSpy, he was drawn to the tales as parables.

In one tale, he was lying in the backseat of his car parked along Mailbox Road when he first saw them: strange spaceships, dotted with lights, hovering. They flew right over the car. It was only later, after thinking about the vivid memories he had had, that he realized he had been lying in a position from which he could not have seen ships overhead. He used the story as an example of how easy it was to delude yourself into thinking you had seen something you had not, how tricky the business of seeing things in the sky near the Black Mailbox was.

Campbell dressed in camou outfits bought at Hahn's Surplus in Las Vegas and talked a lot about his selection of MREs—"meals ready to eat," latter-day K-rations. Sometimes he suggested a grown-up version of the kid in your neighborhood who wanted to play soldiers all the time.

He said that he was not a UFO buff or Stealth fan but a philosopher and inquirer into the nature of truth. His business card read, "Area 51 Research Center. UFOs—Gov't Secrets—Philosophy—Psychology etc." He would make himself the chief researcher into Area 51, an advocate against secrecy, an extremely useful talking head for television crews, and a spy. "I am spying," he would say, "on behalf of the American people."

He had quickly run afoul of Joe and Pat Travis and been exiled to a trailer at the other end of town. The story going around was that Joe, drunk one night, had burst into the trailer and put a gun to Glenn's head. Campbell's story was that Joe had come in "in a drunken rage" and accused him of killing the Inn's business.

Before I came to Nevada, Steve Douglass had told me about PsychoSpy. I phoned and Campbell sent me a copy of his *Area 51 Viewer's Guide*. It compared well with the better travel guides to Europe. The very idea that someone had created a travel guide to a place that did not officially exist was exquisitely appealing to all of us fascinated by Dreamland. In his write-up of the *Guide* for the Federation of American Scientists' Secrecy and Government Bulletin, Steve Aftergood praised "its deliberate epistemological murkiness."

"Don't believe everything you hear," Campbell scrawled on the cover of the copy he sent me. Did I seem naïve? Had he marked me as a skeptic? It was the first sign of his tendency—useful for dealing with the press—to chose his words carefully to match his audience.

I first met him at the trailhead to Freedom Ridge. He was standing beside his battered subcompact car, still with Massachusetts plates and covered with dozens of little stickers from places he had visited. We gave each other a look. I think his was suspicious. I know mine was. I wanted to like him, but from that first moment at the trailhead I had found it difficult to warm to him. The Minister, another Interceptor, suggested that he was trying to overcome his basic shyness in that diffident way shy people often assume. He often had the air of a hired guide, the park ranger of Dreamland, but in time I came to think of him as a jester and as, in that grand American locution, a gadfly. A philosopher, a naturalist of the unnatural, he sometimes suggested a parodic Thoreau, with Groom Dry Lake as his Walden Pond.

If you were simplifying the story of Dreamland for a TV movie, and needed to combine all the Interceptors into one character, Campbell might be that character. He gave good sound bites to the visiting syndicated TV shows, varying his tone, patiently doing retakes. In PsychoSpy, producers could imagine all the Interceptors' modes and dreams wrapped up in this one guy.

Campbell had established what he called "the Whitesides Defense Council," to fight the military's takeover of the viewpoints there and on Freedom Ridge, and then "the Secrecy Oversight Council." He skillfully tapped into the skepticism toward the federal government, one of the few common bonds among Nevadans, and established a middle ground between political right and left, and between the youfers and stealthies who wanted to break down the walls of what they deemed unnecessary government secrecy. He named the ridge he had discovered Freedom Ridge, because who could be against freedom?

Nevadans have long resented federal ownership of the vast majority of their lands, even though the feds pump billions of dollars into the state economy through the military, the DOE, and other agencies. What Mark Twain wrote in *Roughing It* about nineteenth-century Nevadans pretty much holds true today. When they finally achieved statehood, Twain said, "The people were glad to have a legitimately constituted government, but did not particularly enjoy having strangers from distant states put in authority over them—a sentiment that was natural enough."

Soon Campbell began distributing a newsletter called *The Desert Rat* by mail and on the Net. He reported on hearings, arrests of perimeter crossers, and Area 51 rumor and lore. He told of strange characters that appeared in Rachel, like Ambassador Merlyn Merlin II from the planet Draconis, who said he was a "being of Light" (although, PsychoSpy editorialized, "we touched him and found him to be quite solid"). Merlin was on a mission to promote the coming "Golden Age," when the aliens would be integrated into our society and we humans would evolve into a higher form. Campbell came up with an odd source from within the test site: Jarod, who claimed to have worked on "flight simulators" for the alien craft hidden at S-4.

Campbell kept showing up at hearings and working the media, irritating the hell out of the military and the Lincoln County sheriff's department. And he fought the BLM's efforts to seize more land by demanding that the Air Force explain why it needed the land, what it was doing in Dreamland. He bent a classic bureaucratic catch-22 back on itself: The government needed the land to keep secret what it was doing on the base, but because what it was doing was secret it could not explain why it needed the land.

Campbell rightly understood that the hardest thing for the military to deal with was derision. His best line was his response to the signs on the perimeter and the apparatus of secrecy they stood for: "Use of deadly farce authorized."

―――

Glenn Campbell and Jim Goodall designed their own version of a uniform patch for the workers at Groom. It portrayed an Aurora-like craft sweeping up from a dry lake with mountains behind it, and the words "Dreamland Groom Lake Test Site." Soon we heard the patch was for sale at a store in the Pentagon mall. Glenn did not hesitate to print it up on T-shirts and caps and sell it through his catalog and in his "Research Center." Then it was reproduced in several magazines as if it were official, as if it were actually worn by the camou dudes and others inside the base.

―――

A year or so later I was in Nevada again. I phoned Glenn from Las Vegas. "Well," he said, "I'll have to be in the Research Center all day." This tone of pressing business was new.

The Research Center was Glenn's trailer, at the south end of Rachel, and I knew it well. "I've never seen the Research Center," I said, trying to keep the archness out of my voice.

When I pulled up to the little trailer, I noticed junk outside—a cow skull, with bits of skin and hair still clinging to it, and some old aircraft parts. Inside on the ceiling was a large map of Dreamland and surrounding airspace, and a big new Macintosh sat on a table. On the wall was a quotation from anthropologist Margaret Mead about how a few people with conviction can change the world. On the floor were strewn old socks.

Secret aircraft interested Campbell barely at all. He had once said that if the legendary Aurora landed in front of him, taxied up, and ran over his foot, he would pay it no attention.

He handed me an article called "Effects of UFOs upon Human Beings." It dealt with odd electrical effects—radio static, flashing lights—such as those seen in the pickup truck scene in *Close Encounters*. I noticed that the author was named "McCampbell," and the similarity made me wonder if this guy was not some kind of doppelgänger of Glenn's, what he wished he could be—if only he could believe. It was a measured, mostly scientific report on radio interference, sunburn effects, electrical shorts, and other phenomena reported by those who had encountered UFOs.

There were those who believed Campbell was a closet youfer, but he became increasingly skeptical of Lazar. Gene Huff, Lazar's pal, took to calling Campbell "Goober" on-line. "He tends to alienate people," said Huff. "He's a strange bird, a weird guy. He told me he moved out here because of the Bob Lazar story, and now he attacks Bob. I call his operation the UFO division of the Mickey Mouse Club. It may be fine for the shitkickers and dickheads in Rachel, but not for the rest of the world."

Campbell had alienated Huff by publishing transcripts of Lazar's statements at the Ultimate UFO Conference in Rachel, when Huff believed that he was the only one with the right to do so. Huff took a kind of pride in defending Lazar. He made Campbell a particular target, hinting darkly of immoral, even criminal behavior in his background. The word among the Interceptors was that each had something on the other. Campbell had negative information on Lazar's credibility; Huff had info on Campbell. It was a parody of MAD, mutual assured destruction, so neither could use the stories, even if they were true. "Well," Steve

Douglass remarked wryly when he heard of these supposed skeletons, "I guess we all have our little Groom Lakes."

————

Campbell made life difficult for the Air Force by challenging its takeover of additional land, including the Freedom Ridge overlooks. But his whole media act as "searcher for the nature of truth" was as pretentious as his dressing in camou gear and eating MREs.

On one trip to Tonopah, PsychoSpy hurled himself up against the chain-link fence and the guard quickly dropped to a crouch and focused his M-16. Usually you could josh with the guards, but all of sudden the casualness was gone. Everyone in the party was shocked.

The guard had grown increasingly irritated. On their radios, the dudes referred to Campbell as "our friend" or "the editor." Inevitably, he managed to get himself arrested. Accompanying yet another reporter and camera crew, he led the dudes on a chase that ended with the group pinned down by a sandblasting chopper, then cut off by one of the dude mobiles. When they asked for the film, PsychoSpy locked the doors of his Toyota. They finally forced him out and confiscated his film.

He went to court, serving as his own attorney in a thirty-eight-dollar suit he bought at a Mormon thrift shop in Las Vegas. He argued that by not returning his film the dudes were in effect concealing evidence of a crime: They had flown the helicopter that sandblasted him below the five-hundred-foot FAA-mandated minimum altitude.

It was all to no avail, as he knew it would be, and he was fined. His community service included working on a history of Rachel and helping out at the senior center.

————

After Freedom Ridge was closed off in the spring of 1995, and perhaps finding his welcome in Rachel wearing thin, PsychoSpy rented an apartment in a complex in Las Vegas, near the airport. From his front window he could see the Janet flights taking off and landing on their way to and from Groom. He called this his "Las Vegas Research Center." It was a characteristically defiant gesture.

If he sometimes seemed smug, feeding off his appearances in the media even as he spoke disdainfully of its operations, he was better than anyone at cutting to key points about the base. Wrapping himself in his cloak of citizen advocate, he argued that the importance of the Lazar story was not the existence or nonexistence of UFOs in government hands. What mattered was that there could be. Policies of secrecy had made it possible, and those policies were in defiance of all-American moral law and tradition.

PsychoSpy wrote that he—except he used the editorial or royal "we"—approached aliens-at-the-test-site stories "as folklore." "Rather than assuming a story is false until proven true," he stated in The Desert Rat, "we proceed as though it were true, collecting information about it until we reach an insurmountable roadblock or inconsistency." No lie, he was confident, could "reproduce all the rich interconnections of reality.

"As long as a story remains interesting in itself, like a well-constructed novel, we are willing to set aside the issue of truth and go along for the ride," he added, sounding somewhat like Jung himself.

The real problem was that the military and the government—or whatever conspiracy you wanted to postulate—had been able to establish enough secrecy to make the possibility of flying saucers at Area 51 vital for thousands of citizens. At Groom Lake, airplanes had been chopped up, burned, and buried; entire programs were erased from the record. They had kept the first spy satellite secret for more than thirty years and the National Reconnaissance Office, the organization that operated it, unacknowledged and virtually unknown for even longer. They had managed to hide the NRO's "stealth building"—only feet from the strip mall and spec office parks around Dulles Airport—from Congress itself. They had kept secret for years the post-nuclear-war presidential redoubt in the basement of the Greenbrier Hotel in West Virginia. Excessive secrecy left the way open for—fairly demanded—all kinds of conspiratorial speculations. If nothing was seen, much would be dreamed.

———

Lurking on the Internet, as on a high-elevation viewpoint, I saw Dreamland taking on a new, shadowy presence in cyberspace. Sometimes active pilots would appear. There were B-1B crews, chattering and bragging. A few days later they suddenly disappeared. I had the very firm impression that a higher-up had spotted the postings. The B-1 is known to airmen as "the Bone" and its crews, by extension, as "Bonemen," if not "Boners," men of camaraderie and enthusiasm who write poetry "on beer drinking, cannibalism, and such." Perhaps the colonel was not pleased to read about a near-supersonic flight with live ordnance, as in a message headed "Lots o' Iron": "Yesterday, we were 500AGL, .998 Mach, very very near civilized establishments en route to the ft Sill IP with 84; live eggs on board—that, my friends, is the sound of freedom!"

Soon enough, there were not only newsgroups about Dreamland, like alt.conspiracy.area51, but websites devoted to it. Glenn Campbell had previously learned that some mail to Area 51 was directed to "Pittman Station, Henderson." Henderson is a town just east of Las Vegas, the site of a defunct post office that once received mail for the base. One buff plugged "Pittman Station" into the Alta Vista web searcher. It came back with a 1990 NASA press release listing astronaut candidates. Pittman Station was listed as the site of employment of a Capt. Carl E. Walz. Another buff then did a search for "Walz" and came up with a detailed NASA biography. You could read that Walz's parents lived in South Euclid, Ohio, and that he had graduated from Charles T. Brush High School in Lyndhurst, Ohio. You could also find that "in July 1987 he was transferred to Las Vegas, Nevada, where he served as a Flight Test Program Manager at Detachment 3, Air Force Flight Test Center." The Air Force Flight Test Center is located at Edwards Air Force Base, apparently with a detached unit at Groom Lake, with a Pittman Station mail drop: It appeared that Walz had worked in Dreamland. Had he flown Aurora? Been a test pilot for some other project? None of that was answered.

Glenn drew it all together. Having previously established that AFFTC "has a presence at Groom . . . now we know that it is Detachment 3 that is housed there. This is consistent with the designation on the cover of the Area 51 Security Manual of 'DET 3 SP,' with 'SP' perhaps referring to 'Security Police.' "

You could magnify a little detail into a live connection. But you could also magnify rumor—and there were certainly frauds and weird entries. It was not always easy to recognize them, and even when you did, sometimes they were more interesting than the accurate postings.

One man described his "grandpa," who had worked, he said, at Area 51 or at Tonopah, he wasn't sure which. The grandfather would never discuss his work, but when he was dying and had been "given Morphone and other asiditives [sic]," he finally talked. He had been given a "metal" for his work. After his death, "an onslought of Military personnel took the Metal, the Bodys and the licke" away.

Another posting offered a chronology of runway expansions at the base that read like a parody of Tom Mahood's painstaking chronologies of events at Groom or his biography of Bob Lazar. The name of the poster was suspicious to begin with: "Robert Harry Hover." Is that "hover," as in the way a saucer hangs?

The detailed listing buzzed with numbers: runway lengths and elevations, magnetic bearings in degrees, minutes, and seconds. The startling and suggestive things are slipped in between the numbers so that you almost don't notice their implications: "Only 70 Base Personnel knew of this place." "1964 Anti-gravitation device test. Unsuccessful," and the Delphi "1970 Occurrence Friday, September 11, at 10 PM for one-half hour."

A British UFO magazine published a photograph its editors believed depicted the Aurora refueling over the North Sea in formation with F-111s. Steve Douglass was suspicious. He made some phone calls, checked details of the aircraft types, even the engines, and proved it a hoax. But the hoax was intentional: The picture had been produced by Bill Rose, an astronomer and photographer, for the English magazine *Astronomy Now*, specifically to demonstrate how easy it was to fake photos of UFOs or secret airplanes. The UFO magazine had taken it from *Astronomy Now* without permission and, in effect, proved Rose's point.

Another time, Steve received an anonymous letter containing what the sender said were images of two hypersonic aircraft prototypes. The fuzzy photocopy showed two fighter-size aircraft said to be code-named "SANTA." He discovered that the picture actually showed two prototype miniature deep-diving submarines designed and built by an oceanographer named Graham Hawkes.

The more cases like this you read about, the more time you spend on the perimeter, the more you tend to believe in the native human tendency to exaggerate, embroider, and deceive. There were apparently more nuts than were dreamed of in my philosophy at least. Dreamland expanded one's sense of the native human tendency to duplicity—and to spite.

———

One morning, a group of the Interceptors met near Indian Springs. The day's hike was to the top of Mount Stirling, one of the few places from which you can glimpse—albeit from forty or fifty miles away—Papoose Lake and Lazar's "S-4," ostensible site of saucer test flights and hangars hidden inside cliffs.

At the base of the trail, I met the Swiss Mountain Bat, the most distantly based of the Interceptors (he is in fact Swiss), in his rented Ford Explorer. The Bat had read a book called *Above Top Secret,* which recounted Lazar's and other tales of secret facilities cut into cliffs and mountains and deserts in the western United States. The Bat did not find it so hard to believe in hangars inside cliffs at S-4, with door panels camouflaged as rock. An American might be dubious, but the Bat had done his three years in the Swiss military and seen all the underground hangars and command centers in a country whose laws require the provision of a bomb shelter beneath every newly constructed building. "We've done that all over the country," cut into the mountains "like Swiss cheese," he said with a big smile as we bounced up the Forest Service road.

Sure enough, when he got back to Switzerland, the Bat mailed me photographs of some of the Swiss installations. They appeared as rocky cliffs at first glance, but on closer viewing you could pick up perceptible lines, as if discerning a servant's door in the library of a gentleman's mansion.

He worked for an insurance company, but he seemed to live from one visit to Dreamland to another. He sold his photos of S-4 and Groom Lake to the UFO magazines. He appeared in many of the pictures and would show up on the covers of UFO magazines grinning behind his big camera lens and binoculars, in full camouflage. He regularly e-mailed the Interceptors back in the States from the Bat HQ, as he called his home. But he had another agenda, which was to prove that his country of chocolate bars and gold bars, clocks and banks and burghers, had UFOs too, and cover-ups.

"It's no longer about chocolate and cheese in our tiny li'l country now," he crowed in his e-mails. "Very strange things go bump in deep night over here, too." He told of mystery radar blips and lights.

The Explorer bounced up the narrow road, but the hump between tracks was rising with the elevation. Soon we were almost straddling it. When we pulled to a stop, the sweet smell of burning sage and pinyon came from beneath the vehicle where plants had been scorched by the hot exhaust.

During the steep climb, we kept our eyes on the ground, on droppings of various animals, on red lichens—the same color as Lazar's element 115. PsychoSpy, wearing a porkpie hat from Kmart, breathed heavily. Tom Mahood, his eyes deep-set beneath an Aussie hat, led the

way. "If they were serious," he said as the vistas began to open occasionally through the trees, "Lazar and Huff should have come up here before
they went public." They could have brought a video camera and gotten
some proof of the flights, taped the rising saucers. Now even Lazar
thought the saucers had been moved.

The top came suddenly. A turn and then a sweeping view from 8,200
feet; framed by pine boughs with their beads of sap visible like fresh
rain, the lakes slid among the distant overlapping ranges. "There it is,"
Mahood said. "The most secret place in America."

We sat and recovered and ate. I noticed with a start that the bags of
trail mix and peanuts had swollen into little pillows at over 8,000 feet,
transparent balloons.

No one really believed he would see flying saucers, or even hangar
doors. Nor were we here to disprove anything. It was the possibility that
justified the trip. Possibility expanded our credulity.

Maps were consulted. Mahood had marked the borders of the test
site and range in fluorescent pink and orange and, with his engineer's
eye, had carefully worked out the sight lines from Mount Stirling to
Papoose Lake and other vistas. He squatted on a folding canvas-and-
metal stool and aligned his telescope.

The test site spread out below: the geometric assemblage of Mercury,
the company town where I remembered seeing ads in the cafeteria for
bowling leagues. Highway 95 showed the silver slugs of trucks and the
black flecks of cars. Near the entrance to the test site, you could see the
holding pen where so many demonstrators had been sequestered over
the years. To the east were the Ranger Mountains and a strip of public
land. On an earlier trip, the Interceptors had tried to get up to the edge
of the test site that way, but it was rougher and longer than it had looked,
and some of them nearly collapsed for lack of water.

You could see the stubs of towers where atomic blasts had been set
off, Yucca Flat, Frenchman Flat, the barely visible Command Post and assembly buildings. I imagined what it would have been like to watch the
atmospheric tests from up here, to see the mushroom clouds rising from
the plains, to feel—half a minute, forty-five seconds later—the vast wave
of shock and heat there amid the pines.

But our attention was focused on the light strip of Papoose Lake. We
should have been looking straight at the wall with the sand-colored
doors of secret saucer hangars, pitched at a 30-degree angle, that Lazar
had talked of. As the sun moved lower, the light molded the hills into

fuller shapes. The landscape seemed to puff up and grow fuller, more sculpted, as if inflating like our little bags of food.

"That would be the way to walk in," Mahood said. "Up Nye Canyon, across Frenchman Flat."

To the far right you could see the airfield at Indian Springs with its little x of runways half-flattened like a folding chair from the perspective. We saw a light aircraft pass by. Then someone caught sight of what looked like a building, a blockhouse-like structure, beyond the edge of the test site.

"As the afternoon wore on, they could see more and more," someone intoned in the voice of a television narrator. The whole experience of Interceptordom was cast by the way they figured in TV interviews.

The Swiss Mountain Bat jumped. With the light shifting, he could see the mystery building in his viewfinder. He steadied the huge telephoto lens and delightedly snapped away.

The shadows lengthened. All of a sudden a dark shape came hurtling down from the right of our camp: a great bird, concentrating on its prey, surprised to find us here. I could see its stunned look, and as it pulled up then banked away down the slope, the bird—it was a golden eagle—lost a single white feather. The down feather drifted slowly, like a parachute flare, until it landed in the bushes, and Mahood scrambled to retrieve it.

———

That night, back on the plain beneath Mount Stirling, we camped and built a fire of the pitiful gatherings of twisted driftwood-like pinyon and the odd two-by-four someone had left—and talked of Lazar's shady past and proton cannons and skyquakes. The campfire carved out a cave of light, and lore and jokes flew back and forth as the mesquite burned longer than the thin, twisted sticks seemed to have a right to do. This was the closest we got, I figured, to ghost stories or the primal folktale.

Tom Mahood talked of a guy who claimed to have worked on flying saucer simulators, of a man who knew the man who did the ejection seat for the Blackbird and then for the Aurora. "My rational mind says that what you see is all there is," he said, "but another part of me wants to believe there's something else out there."

This from an engineer, an exacting thinker and researcher—yet from the beginning Mahood was drawn to Lazar's story. For all his just-the-facts-ma'am attitude, he admitted he had initially felt that Lazar was telling the truth, that he had been at S-4. Lazar's very presence seemed to have this convincing effect on people, his sense of self-possession, almost diffidence. Mahood continued to feel that way even as more and more information seemed to discredit Lazar's claims about his résumé and career. But the emotional link remained. For a time, it gave him restless nights. Like many, Mahood wanted to believe; the wild hopes and cosmic dreams of his heart struggled with his engineer's head. As if in a Pascalian leap into faith, even the most remote of chances that the revelations were true provided a tempting counterbalance to the weight of facts against them.

———

The darkness was profound. I lay on my back in the sleeping bag and saw the Milky Way not as a collection of stars but as a smear. My eye went to Orion, the first constellation I had learned at the planetarium in second grade. Zeta Reticuli, putative source of flying saucers, was visible in Cygnus, someone had told me, but only in the Southern Hemisphere. All at once a single meteorite streaked through the Little Dipper. I wouldn't have believed it if I hadn't seen it myself.

Driving back to Las Vegas the next morning, I thought about the shadows around the campfire. They suggested to me Plato's cave, as updated by the class cut-up who shapes his fingers into a rabbit or duck by flashing them in the beam of the film-strip projector. Ordinary realities, kept out of sight, turn into flickering monsters on the cave wall. Captured foreign fighters become alien spaceships; nuclear test tunnels turn into a network of secret underground chambers; radar test shapes on stalks are transformed into saucers hovering above apertures to the underworld.

A few weeks after the hike, Tom Mahood returned to the top of Mount Stirling. With a different telescope, by different light, he was sure that the mysterious building we had seen was nothing more than a rock formation.

14. Black Manta

Dreamland spun me out again, this time to Amarillo, home of the famed Cadillac Ranch, where old Caddies are buried up to their tail fins, and whose name always reminds me of "Paradise Ranch." Tracking the evolution of the tail fin, which was inspired by the fins on Kelly Johnson's P-38 and the rocket fins of the late fifties, I understood it as a monument to a society in which El Dorado is no longer a mythical gold city but a Cadillac coupe with a vinyl top and gold-anodized brightwork installed at the dealership.

I came here to meet Steve Douglass. Although he rarely ventured to the perimeter of Dreamland, he came to be venerated as the ur-Interceptor, a near legendary figure. He was a military monitor, an interceptor of radio broadcasts, and if anyone knew what was flying, he did.

Beavis and Butt-head snickered on the TV in the living room of Steve's ranch-style home. Then Steve popped a tape into the VCR and the boys disappeared into a powdery mix of colors. There was a silence, then solid gray-blue; then a dot emerged, grew larger, became a bat, a ray-shaped airplane swooping overhead—and finally the image dissolved into gray grit. Steve flicked the set off. "Seven seconds," he said. "You live for those moments. You listen all those hours for that kind of gold nugget."

Steve felt sure he had captured the TR3A Black Manta on video for the very first time.

———

Once a year, something like Brigadoon, the long-closed base at Roswell came alive with dozens of aircraft, in April or May, on the occasion of the annual Roving Sands air exercise, which involves hundreds of airplanes and airspace over five states. For Steve and, later, other Interceptors, Roving Sands was the Olympics of plane spotting.

Trying to spot something unusual meant spending hours standing in the back of a pickup truck, listening to scanners and watching the distant horizons. Most of the planes were familiar ones, but every so often something strange would appear, usually at dusk. In May 1993, Steve had gone black-plane hunting there with Elwood Johnston, his father-in-law and fellow stealthy. Steve had always had good luck finding mysterious flying objects when Elwood was along; he believed Elwood was a stealth lure.

At the end of the day, Elwood saw it first. "What's that?" Then they both found it on the horizon in the dusk. He was sure it was not an F-117; it was slower, with a different sound, a different shape. Douglass's radio scanner crackled, the numbers churned on its readout. As he raised his video camera, the battery warning light flashed. He grabbed seven seconds of video before the machine snapped off.

When he got home, Douglass printed an enhanced view of the bat plane. Then, consulting with his wide network of experts in the industry, the aviation press, and the military, he tweaked the details to create a speculative image of the airplane.

It looked like the airplane that the Greenpeace intruders had spotted when they ventured across the Dreamland perimeter in 1986. It had been speculated on as far back as 1990 by the mysterious figure who signed himself "J. Jones," an insider reporting on stealth. The accepted wisdom among stealth chasers and Interceptors was that the Black Manta operated in tandem with the F-117A Stealth fighter, relaying target information, and evidence suggested it had been used in the Gulf War.

The TR3A would likely have first flown from Groom Lake, and something like it had been seen often near Dreamland. In 1993, Agent X had been startled to see a batlike craft sailing right over the highway near Alamo, a little town of two motels and a diner southeast of the perimeter. Even the radio host Art Bell had spotted a two-hundred-foot hovering triangle near his home in Pahrump, west of Dreamland. And what about the two sightings, west and east of the restricted area, of triangular craft in company with F-117s, in February and November of 1995?

Was the TR3A a descendant of the Theater High Altitude Penetrator (THAP), an airplane made public by Northrop in the late 1970s? In 1980 both THAP—which seemed to be Northrop's consolation prize in the original stealth competition—and the Lockheed Have Blue programs were classified Special Access Programs, and very little additional information about them emerged. But in 1983, defense industry insiders reported that Northrop had gotten a contract for twenty-five or so aircraft. In November 1987, Steve learned of a crash of a plane "not an F-117."

Was it the same craft involved in the September 26, 1994, Boscombe Down crash in the United Kingdom? Was it the craft that crashed on October 18, 1994, at Kirtland AFB? Steve picked up the radio traffic from the crash recovery team, which referred to returning the wreckage of "a high-altitude research aircraft" to Edwards.

Douglass's printout of the TR3A looks, at first glance, like a flying saucer.

———

Steve leads me into his thickly carpeted retreat, where six scanners work steadily, hopping from channel to channel—shortwave, VHF, UHF, sideband—all feeding into a little voice-activated Radio Shack tape recorder that vacuums up every scrap of voice, packing a day's talk into ninety minutes or so, which Douglass listens to late at night. He grows restless without a scanner nearby, the bubbling reassurance of its red digits pumping frequencies through its chips and extruding slugs of conversation. Talk with him on the phone, and you have to get used to sudden soft pauses, as if there were a fault in the line, as he cocks his other ear toward something on one of the radios.

Airplane models hang from the ceiling; pictures of planes line the walls. In one corner lurks a huge oscilloscope—military surplus—and a Hallicrafter's shortwave set, packed with tubes, picked up for twenty-five dollars at a garage sale. There are maps of military bases and of New Mexico, as well as a Landsat photo of the F-117 base at Tonopah. Red and blue lines show main air routes and refueling courses. Amarillo is dead center in the heart of the country's military flyways. "Why go to Groom Lake," Steve asks, "when the planes seem to come to you?" Steve has had Stealth fighters fly right over the house.

Steve loads the taped radio clips from the White Sands episode onto Soundscan files on his Performa 450 computer. Now he clicks on each little folder on the screen. Maintenance and security people talk about the arrival of a VIP in the morning. (Later, Douglass learns that Gen. Colin Powell, then chair of the Joint Chiefs of Staff, had visited El Paso the day before. He suspects Powell might have made an unpublicized side trip for a glimpse of the Black Manta.) The radio traffic refers to the plane as an "STF." Does that stand for Stealth or Survivable Tactical Fighter? Another report has it as "Tactical Survivable Aircraft" and ties its lineage to the Northrop THAP of earlier years.

At the history office at Edwards Air Force Base, I find that the TR3A is included among the aircraft types index, along with the F-15 and the SR-71. But the only materials in the file are press clips, not the flight test reports and other documents provided for other aircraft.

———

On March 23, 1992, Steve picked up radio traffic between two unidentified craft using the call signs Darkstar November and Darkstar Mike. He realized they were flying close by. Then the house began to shake, and the window frames rattled. He ran out of his house, slapping film into his Canon AE-1. He could hear the rumbling sound of the engine—like a rocket engine, only intermittent—in regular bursts. He could even feel it in his chest, but all he saw of the craft itself was "a silver glint of light, a metallic shape." Even with a 400mm telephoto lens he managed to capture only the plane's contrail—a string of roundish puffs, the now-well-known "doughnut-on-a-rope."

About a month later, he got a report from one of his Interceptor informants in the high desert area of California of communications between "Joshua control" and an aircraft calling itself "Gaspipe." He and other watchers thought it was the spoor of a new kind of engine, a secret aircraft's pulser jet. Within weeks it was published in *Aviation Week*.

Later, Steve talked by phone with a pulse-jet engine expert he knows at a military contractor. The engineer played chords on a synthesizer over the phone, striking lower and lower frequencies until Douglass found the one he had heard. "Damn," the engineer said, recognizing that his rivals had perfected an advanced jet engine, "they've done it."

———

Steve Douglass grew up in the West, and one of his earliest memories is of going up into the mountains to watch in the distance a nuclear explosion in Nevada. He recalls how beautiful it was, how it lit up the sky like a rosy sunset.

He was a photographer for the newspaper in Amarillo when he bought his first scanner. It was a simple model for following the police and fire bands, so he could rush to the scene of a car accident or warehouse conflagration to snap pictures for the paper. The more he listened, the more he wondered what else was on the air. "It was like the old George Carlin bit," he says, " 'What's on beyond the edge of the dial, after the knob stops?' What are they hiding out there?"

What he discovered was the new world of scanners. Around 1970, solid-state electronics had replaced old crystals as the heart of scanners. Before long, you could buy a 200-channel scanner from Radio Shack for about three hundred dollars. Radio Shack has sold more than four million 2006 scanners worldwide, and in theory anyone who knows how to use one can eavesdrop on most military traffic, on *Air Force One* itself. Steve figured there were probably no more than five hundred hard-core military monitors in this country, which may simply mean people who have nothing better to do with their time.

Some systems hop from channel to channel to defeat eavesdroppers, but the best of the new equipment can cover thousands of channels a second and listen in on the channel-hoppers too. Encryption is used at high-level bases, but it's expensive and vulnerable to atmospheric shifts. Even at Groom Lake, the camou dudes broadcast in the open, most of the time.

Steve bought more powerful scanners and eventually found all sorts of strange things beyond the end of the dial. He began picking up the military channels, and as a stringer he fed bits of information to the Associated Press. His first scoop came in 1986, when he picked up transmissions from a Soviet nuclear sub with a critical nuclear reactor problem. In an early sign of détente, U.S. Navy ships had rushed to the scene to help out. The Pentagon denied the story, but when an A.P. reporter brought in Douglass's tape, on which a sailor screams, "It's sinking! It's going down! Radiation counters are going up!" the military finally admitted what was going on. Television cameras were present when American ships rescued the Soviet crew.

Douglass heard the troops assembling to invade Grenada, then Panama. During the Gulf War, he fed network reporters shortwave accounts of SCUD launchings from troops in Saudi Arabia before their Israeli bureaus heard the sirens.

Once, he monitored the radio traffic surrounding the crash of a B-1B bomber. The airplane had smashed into a mountainside, and the Air

Force was blaming the pilot. Investigators turned up at the pilot's father's house, asking if his son was a homosexual or a drug abuser. Congress was considering further funding for the B-1B, and the Air Force wanted nothing said that could pin fault on the airplane. Douglass, after hearing of the crash, checked his scanner tape from the night before. It clearly recorded the pilot complaining of problems with the plane's autopilot. When the news came out, the military brass denied that any "amateur ham-radio operator" could have such information. The father of the dead pilot called Steve, and Steve was able to tell him the truth.

———

Steve and his wife, Teresa, an artist and computer whiz, began to publish a newsletter for military monitors called *Intercepts*. It was the first publication that tied together the far-flung watchers, who were conscious of being, well, outside of the mainstream, and it gave them a sense of the others out there. In the pages of *Intercepts* Douglass ran letters and columns above the code names of correspondents: Darkstar November, Big Red, Lone Star, Ghostrider. Douglass soon had subscribers at CIA headquarters in Langley, Virginia, and throughout the military.

He established his own dial-up bulletin board, which bubbled away on an old Commodore 128 in his Bat Cave, and then began operating a forum on America Online called Above Top Secret, later Project Black. His book, *The Comprehensive Guide to Military Monitoring*, would become the bible of the subject in which he shares tricks, frequencies, and some of the wonderful American music of call signs and radio vocabulary.

On the Net, Steve saw that many stealth chasers were meeting each other for the first time. On-line had become a clearinghouse for monitors to share military intelligence—"a public intelligence network," he termed it. Before "It's true, I saw it on the Internet" became a joke, Steve saw clues and rumors blossom on-line.

One day, men in suits began to appear in the windows of the long-vacant house behind Steve and Teresa's place, and a bug showed up on his phone. The cellular phone of the local congressman had been tapped and recorded, and Douglass was suspected. The real culprit was later found, but Douglass now sweeps for bugs monthly.

———

Military monitoring has been called "ham radio cubed" or "super ham," and the scoops monitors provide are often described disdainfully as "ham radio reports," Steve told me. But monitoring is a whole different culture from that of ham radio operators. "Hams look down on monitors," he said. "Hams say, 'You can only listen?' " The monitoring culture is actually more closely related to the nation's brief infatuation in the mid-seventies with CB radio. Motorists listened to CB to find out what the police were doing; many then graduated to police scanners, which let them listen in on the police directly. Steve, too, played around with CB until "the idiots all got on."

Steve established good contacts in the military and defense industries, and he was constantly hanging around bases and watching exercises. Once he chatted with a Stealth fighter pilot at an air show. "How does flying the F-117 compare to the TR3A?" Steve asked him.

"Well, you see . . ." the pilot began. "Huh? Well, I can't talk about that."

———

In 1995, Steve spotted what he thought was another new aircraft near Cannon Air Force Base. He called it "artichoke," for the pointed, overlapping shapes of its trailing edge. It was light gray and about the size of an F-111. He noticed that the military had opened a new Military Operating Area—a MOA—not far from where he had spotted the so-called artichoke. Another year, Steve spotted lights hovering above the runway at the White Sands base, then zipping off—the classic flying saucer flight pattern. But from the radio traffic, he speculated that this might be another secret craft—a Harrier-like vertical-takeoff-and-landing fighter.

Steve noted other sightings, some of them halfway around the world. In September 1994, reports surfaced in the United Kingdom of a mysterious aircraft crash-landing at Boscombe Down, the U.K.'s own version of Dreamland. The British had been testing at least one craft in Dreamland, the evidence suggested, but this was likely an American plane.

The front landing gear, it seemed, had failed, and observers spotted the strange shape, with inward-canting tail fins, partly covered by a tarpaulin. According to one theory it was a spin-off of Northrop's YF-23, which competed unsuccessfully with Lockheed-Boeing's YF-22 to replace the F-15. Some believed it was called ASTRA, for Advanced Stealth Technology Reconnaissance Aircraft, built by Northrop and McDonnell Douglas. Some believed the charcoal gray airplane was the A-17, a replacement for the F-111. Radio intercepts referred to it as AV-6 (Air

Vehicle Six, its construction number), with USAF serial number 90-2414 and the call sign Blackbuck 11.

The C-5 that came to retrieve it was referred to by the call sign "Lance 18," with the intended destination code of KPMD—air controller designation for the airport at Palmdale, California, home of Northrop and Lockheed. Within days, there was another accident involving a similar plane in New Mexico. To Steve that suggested some structural problem: Had the same part failed after the same length of time and stress?

———

In the spring of 1994, I headed out one evening at dusk to watch the Roving Sands exercises in Roswell. Across from the base, in front of a neat little house, a tiny Hispanic woman was tenderly washing her husband's highway patrol car. The landscape on one side of the road was the stuff of sport-utility ads or Technicolor Westerns, with barbed-wire cattle fences and a decaying windmill for punctuation. On the other side, it was action thriller: chain-link fence topped with accordion wire.

While Steve kept his video camera focused upward, I found my eyes wandering to the ground. Even when passing directly overhead, fighters did not seem to be moving at astonishing speed until you stood back and watched the first appearance of their shadows, then the black wave racing toward you across dry grass.

At night, it is harder to see. There is a certain phenomenology with regard to night visions, the green, teeming imagery familiar from crime and war footage, but to the unaided eye, the lights of the most mundane craft could grow surreal. Far off in the Hollywood Western sky, a Navy fighter returning to base was just a grain of red. But then it defined itself, stronger, like a laser, before innocently resolving into two wing-tip lights as it came in to land.

Standing on the fence line, waiting for flying objects we could identity, Steve and I had time to contemplate the appearances of UFOs. "Why don't they come as clouds," I asked, "so they blend in?"

"Why don't they come disguised as Golden Arches," Steve replied, "so they can land wherever they want and not be noticed?"

"What if they're unofficial, not authorized? We always assume they're here doing research, but what if they're not part of some E.T. NASA, but just teenagers out cruising?"

"Joyriders," Steve said. "Heck, low riders."

Sometimes, Steve told me, he used his scanners to pick up tornado sightings and went off chasing them. "You know," he said, "trailer parks cause tornadoes. It's a scientific fact."

"And maybe deserts cause UFOs? How come they never land in the parking lots of New Jersey malls?"

Or was it military bases and secret research facilities that spawned saucers? Were they there, as in the famed reports of snooping UFOs at Malmstrom and Kirtland Air Force bases, to spy on us? Or was it the presence of all that strange and frightening weaponry that put people in a frame of mind to see discs and lights?

———

Glenn Campbell came down for one edition of Roving Sands, and Steve was stunned to hear the security forces speak of him on the radio. "PsychoSpy is here," they said. Campbell thought it was a joke at first, but it startled and perhaps frightened him. That year, too, a SEAL unit was assigned to patrol the Roswell base perimeter. Steve and the others watched a helicopter drop off several shadowy figures. The strange thing wasn't that they were black, but that they were silent. Steve could hear cars on the road a couple of miles away, but no sound from the helicopter engines.

The SEALs crept up on the Interceptors, but they were clearly visible on the Interceptors' night-vision scopes and clearly audible on the scanners. Every time the SEALs moved up a little bit, the Interceptors lit them with a laser. It was Radio Shack versus the Pentagon, and it was no contest.

Finally, the SEALs got disgusted and radioed for a humvee to pick them up. One of the Interceptors had an idea and got on a cell phone. A few minutes later, a Domino's delivery vehicle pulled up to the humvee parked back on the perimeter and delivered two large ones, pepperoni and cheese. The anchovies were the final insult.

———

Just as Bob Lazar's story held a fascination that grew even as it seemed less and less likely to be true, the Roswell story was drawing more attention the further it receded into the past. When Steve and I visited the town, it was on the verge of a new saucer boom.

A mayor was elected who decided to make lemonade from what respectable townspeople had hitherto viewed as a lemon. He put a saucer on his official stationery, and promoted the town as a saucer tourist site.

"Why don't you drop in?" the brochures read. "After all, THEY might have."

James Moseley, the acerbic publisher of *Saucer Smear*, commented in 1996 that "the Roswell incident has emerged as a myth of such power and allure that it is no longer in anybody's best interest to seek—or admit—the truth."

What made the Roswell story special was that it marked the only time the U.S. government ever officially claimed to have captured a flying saucer. The press release from Roswell Army Air Field came on July 8, 1947:

> The many rumors regarding the flying disc became a reality yesterday when the intelligence office of the 509th Bomb Group of the Eighth Air Force, Roswell Army Air Field, was fortunate enough to gain possession of a disc through the cooperation of one of the local ranchers and the sheriff's office of Chaves County. The flying object landed on a ranch near Roswell sometime last week. Not having phone facilities, the rancher stored the wreckage . . .

"Fortunate," "cooperation"—the studied casualness of the vocabulary made the acquisition sound like a bequest to a museum.

The man who issued the press release—on orders from his boss, Gen. William "Butch" Blanchard, head of the 509th Bomb Group—was public information officer Lt. Walter Haut. Haut, interviewed years later, declared he had never seen the wreckage, or visited the crash site. At first, he didn't really think it was a flying saucer, but by the 1980s he did.

The 509th was the special unit that had dropped the first atomic bombs on Hiroshima and Nagasaki in 1945 and, the year before, had bombed Eniwetok. In 1947, it was the only unit in the world capable of dropping atomic weapons, although there were still precious few of them in the U.S. arsenal. Blanchard was a top bombardier who had trained in the secret range near Wendover, Utah. Had his bomb scores been a few points higher, he might have pushed the button himself over Hiroshima.

Haut would attribute Blanchard's apparent eagerness to issue the press release to his long-standing interest in good community relations. Butch Blanchard carefully doled out news from the base evenhandedly between the two newspapers and two radio stations of Roswell. On July 8, 1947, the Roswell *Daily Record*, the local afternoon paper, ran the following story:

RAAF CAPTURES FLYING SAUCER
ON RANCH IN ROSWELL REGION

No Details of Flying Disk Are Revealed
Roswell Hardware Man and Wife Report Disk Seen

The intelligence office of the 509th Bombardment Group at Roswell Army Field announced at noon today that the field has come into possession of a flying saucer. According to information released by the department, over authority of Maj. J. A. Marcel, intelligence officer, the disk was recovered on a ranch in the Roswell vicinity, after an unidentified rancher had notified Sheriff Geo. Wilcox, here, that he had found the instrument on his premises. Major Marcel and a detail from the base went to the ranch and recovered the disk, it was stated. After the intelligence officer here had inspected the instrument it was flown to higher headquarters.

The headlines did not last long. The wreckage was flown to the headquarters of the Eighth Air Force in Fort Worth, where the commanding general, Roger Ramey, declared the "instrument" to be a weather balloon. Only afternoon papers and one round of the eastern morning papers carried the story before the press release was "withdrawn."

Haut was dispatched to retrieve copies of the release from the local newspapers and radio stations, although he denied later reports that he had been ordered to do so directly by the Pentagon. And neither his career nor Blanchard's seem to have suffered for their embarrassing haste.

The New York Times treated the story on its front page on Wednesday, July 9, 1947: 'DISC' NEAR BOMB TEST SITE IS JUST A WEATHER BALLOON—emphasizing the proximity of Alamogordo, a hundred miles away. 'FLYING SAUCER' TALES POUR IN FROM ROUND THE WORLD.

Walter Haut's named was mangled as "Warren Haught," and the story made Irving Newton, the weather man in Ramey's office, the hero for identifying the wreckage as a weather balloon. But there was enough publicity to keep the phones in Roswell ringing for a week or so.

A crestfallen Daily Record the next day reported, RAMEY EMPTIES ROSWELL SAUCER, and HARASSED RANCHER WHO LOCATED "SAUCER" SORRY HE TOLD ABOUT IT. After the affair had faded, the newspaper ran an editorial that implicitly set the whole thing in the context of the wider saucer obsession and the nation's mood with regard to atomic weapons and international tensions. The end of the war was supposed to allow Americans to come home. Instead, the beginning of the Cold War had drawn the country into world issues seemingly without resolution. Victory was no longer a goal.

_____

By the early eighties, Roswell had resurfaced. Now it was not just a sighting, it had a plot—a story. Or it *was* a plot—a cover-up. There had been hundreds of flying saucer sightings in the summer of 1947, but Roswell had the elements of a great one. The tale was appealingly simple and dramatic: A couple of days after a violent thunderstorm, during which he hears another sort of boom amid the thunder, an old sheep rancher comes across mysterious wreckage in one of his fields—metal that won't dent, parchment that won't tear, something like balsa wood that won't break, and mysterious writing. He has no phone. He finally makes it to the feed-and-seed town of Roswell and tells the sheriff, who sends him to the Army Air Field base. Two men come to investigate: an intelligence officer and a counterintelligence man. They spend the night in a shack, dining on cold beans and crackers, then rise at dawn to recover the wreckage. There is too much to fit in their old Buick and Jeep carry-all. Soon, troops appear and the area is cordoned off for full recovery. The rancher goes incommunicado for several days, then returns home taciturn but with a new pickup truck. The Air Force ships plane and bodies to Wright Field in Dayton for analysis, issues a press release, and then withdraws it. The FBI stops a wire-service transmission in midsentence, and other government agents intimidate witnesses until all publicity vanishes.

The scenario was perfect: the mysterious shiny object in the desert, the grizzled and baffled rancher, the swift deployment of the military, the hush-hush ferrying of mysterious wreckage to secret labs, the cover-up. The wreckage bore mysterious symbols or signs or letters—hieroglyphic, they were called, "like Japanese or Chinese." It was made for the movies.

Roswell also offered a wonderful cast of characters, from Jesse Marcel to Mac Brazel; the stock sheriff, George Wilcox; and, among the witnesses who would surface later, Pappy Henderson, the old pilot who gave testimony to his friend the dentist. There was Sheridan Cavitt, the closemouthed CIC agent, and the nurse who attended the autopsies and then vanished—there were rumors of a fatal plane crash, rumors she was in a convent on the outskirts of Roswell. And don't forget the teletype operator named Lydia Sleppy, who described the FBI interrupting her dispatch of the press release.

———

The world of ufology is as racked by jealousy and inbreeding as any academic discipline ever dreamed of being. It has its fads and fashions, its hot areas and its backwaters. UFO researchers, like professors of English Romantic poetry or biochemistry, have to find the right topics if they are to flourish on the lecture tour and in the publishing world.

The specialty called "crash recovery" was, in the seventies, a highly unfashionable and suspect realm of the rivalrous and quarrelsome world of the youfers. It had lain in the shadows since the early fifties, when Frank Scully, author of *Behind the Flying Saucers*, had been duped by a couple of con men into believing he was in possession of flying saucer wreckage, which turned out to be profoundly terrestrial pot-and-pan-grade aluminum.

By the late seventies, a UFO researcher named Len Stringfield had found his own niche of credibility, gathering hard evidence about crashes and lining up witnesses. Soon he was a prominent figure on the UFO lecture circuit. He was the first to focus again on Roswell and the recovered wreckage.

Sensing Stringfield was on to a good thing, William Moore and Charles Berlitz, whose previous success had been a book about the Bermuda Triangle, joined with UFO researcher Stanton Friedman in 1980 to publish *The Roswell Incident*, which brought the story back to the forefront. Books by other investigators followed, each adding witnesses and in some cases new, secondary crash sites. In 1989 the Showtime cable network broadcast a film on the case. Roswell came to be the touchstone of the cover-up theory.

The key new witness to emerge was intelligence officer Jesse Marcel, who before his death in 1992 told Friedman of going out to the crash site, near Corona, about fifty miles north of Roswell, with CIC agent Cavitt and collecting pieces of wreckage.

Marcel would later be photographed in the Fort Worth office of General Ramey with parts of the wreckage laid across the office chairs. Ramey had declared the wreckage to be that of a weather balloon—but Marcel knew it was "not of this world."

A key element of the new version of the story was a second crash site, near St. Augustin, New Mexico. Did General Blanchard use his "leave," beginning July 8, to visit this second crash site? Why did Air Force chief of staff Nathan Twining cancel an important trip to the West Coast on July 8? In 1992, Gen. Thomas DuBose (in 1947 a colonel and Ramey's chief of staff) testified shortly before his death that he had received a telephone call from Gen. Clements McMullen at Andrews Army Air Field in Washington, D.C., instructing Ramey to concoct a "cover story" to "get the press off our backs." In 1990, retired general Arthur E. Exon, who had been stationed at Wright Patterson, described the testing there of the Roswell crash debris. The tests included "everything from chemical analysis, stress tests, compression tests, flexing. It was brought into our material evaluation labs. [Some of it] could be easily ripped or changed . . . there were other parts of it that were very thin but awfully strong and couldn't be dented with heavy hammers . . . the overall consensus was that the pieces were from space."

In 1994, after prodding by New Mexico congressman Steven Schiff, the General Accounting Office was assigned the task of ferreting out the truth about Roswell.

But on September 8, 1994, before the report could be completed, the Air Force issued its first official statement on Roswell in forty-seven years—a twenty-three-page report stating that the "most likely" source of the Roswell debris was a balloon from a secret program known as Project Mogul. The purpose of Project Mogul was to detect Soviet nuclear tests by using sensitive instruments carried aloft by high-altitude balloons.

In July 1995, the General Accounting Office released a report based on its search for any government documents about the Roswell crash. It forthrightly declared, "In our search for records concerning the Roswell crash, we learned that some government records covering RAAF (Roswell Army Air Field) activities had been destroyed and others had not." Exactly which had been destroyed and why was not made clear.

The youfers found it suspicious that not only was there no mention at the time of the incident in the Roswell base newspaper The Atomic Blast, but that the official Air Force investigation into UFO sightings, Project Blue Book, did not mention it either. To them the very absence of information meant the presence of something important.

In the Roswell revival, the old mysteries of "Hangar 18" at Wright-Pat, where the debris had supposedly been taken, expanded into a larger, much vaguer pattern. The newest scenarios had B-29s, B-25s, and C-54s dispersing bodies and wreckage in all directions. The element of the story that would tie Roswell to Dreamland was what I thought of as "the Dispersal." In the initial story, the Air Force said that the wreckage had been taken to Fort Worth, then sent on to Wright-Pat. With retelling and new witnesses, and especially with the addition of bodies, the tale of Wright Patterson Air Force Base evolved into one about that base's Hangar 18, a ghostly facility where many saucers and bodies would be stored. Other delivery sites were added: Los Alamos, Sandia, Edwards, McDill or Eglin Air Force bases in Florida, even Indian Springs—and Area 51. In time, Area 51 would become the equivalent of Hangar 18, writ large.

————

Among the new Roswell witnesses who seemed to show up every few months was a man named Frank Kaufman, who had worked at the base in 1947. Appearing in yet another television documentary about the crash, he was a hard-faced man who described having taken part in the recovery of a saucer that was about twenty-five feet long. It had split open in the crash, and alien bodies were recovered. Kaufman had even submitted an official report, he said, with a drawing of the craft. When I caught sight of it, I jumped: It looked a little like a flying saucer, but it looked even more like Steve's Black Manta.

The plastic model of the saucer released by Testor in time for the fiftieth anniversary of the incident was based on drawings "forensically composited," in the company's phrase, from interviews with Kaufman and others conducted by one William Louis McDonald, who had been brought in by John Andrews. Kaufman described the craft as "Stealth-bomber-like"—an example, the Air Force would have said, of the seepage of subsequent impressions into earlier memories that figured in Air Force rebuttals of Roswell witnesses.

With an elongated, raylike body whose edges curled up to form fins or tail, the model craft suggested a crossbreed between saucer and secret airplane, a bastard offspring of the two watcher cultures.

As the Roswell myth grew—and while two academic anthropologists, Benson Saler and Charles Ziegler, analyzed its different versions with the thoroughness of a Lévi-Strauss or Franz Boas—the dispersal of the wreckage traced a direct line in the Lore from Roswell to Dreamland, a line that began as a faint thread of rumor in the eighties and grew to the bold stroke of legend in the nineties.

15. "Redlight" and "MJ"

The road from Roswell to Dreamland was long and circuitous, paved with much speculation and mysterious code names. The name "Redlight," stenciled on shipping crates, marked the first important public connection of Area 51 to flying saucers.

In April 1980 a witness named "Mike," who claimed to be a former employee at Groom Lake, told a representative of MUFON, one of the larger and most influential UFO organizations, that he had caught a glimpse of the crates when he worked at the base between 1961 and 1963. Later, he saw a flying saucer and learned that Project Redlight was the name of a secret program for testing the saucers that had continued until 1962. Mike's story was in part corroborated by radar operators at Tonopah, who around the same time had reported seeing very fast blips on their screens.

The A-12—the original Blackbird, the CIA's version of the airplane—had first flown in 1962. Had Mike only reported the inevitable result of secrecy, the scuttlebutt that grew out of compartmentalized information, the speculations inspired by a few astoundingly real details of a very fast aircraft?

The first time the American public at large ever heard the word "Dreamland" was on the evening of October 14, 1988, when it was uttered on the Fox network's television program *UFO Cover-up? Live!* Two "inside informants" were presented in disguise, code-named Falcon and Condor. Condor described Project Aquarius, an effort to make contact with extraterrestrials, and Snowbird, another program begun in 1972 and still being carried out in Nevada at "an area called Area 51, or Dreamland." "The extraterrestrials," Condor said, "have complete control of this base," the result of an agreement between the government and the aliens that had gone awry. And in what became the most memorable and derided phrase of the show, Condor went on to describe the aliens here on earth, saying, "they enjoy music . . . especially ancient Tibetan–style music . . . the favorite dish or snack is ice cream— especially strawberry."

Condor, many in the UFO world concluded, was actually a man named Richard Doty, a special agent at the Kirtland AFB unit of the Air Force Office of Special Investigations (AFOSI). In September 1980, Doty had written a report concerning a series of unidentified lights seen over the Kirtland range, near Albuquerque and the Manzano Nuclear Weapons Storage Facility.

He was aware of a man named Paul Bennewitz, who believed that aliens were not only actively flying over the range but had implanted human abductees with control devices. Bennewitz claimed to have electronically picked up the signals that activated these devices and to be in communication with aliens inside their craft. Doty was said to have provided him with disinformation, encouraging his speculations, including a memo analyzing Bennewitz's sightings, which seemed to lend them credence—and prove that AFOSI was watching the saucer watchers as well.

Bennewitz, a mild-mannered Albuquerque businessman, would end up chain-smoking and sleepless, with knives and guns stashed around his house, fearing spies and intruders. Finally, he had to be hospitalized for exhaustion.

One piece of disinformation shown to Bennewitz was a faked memo ostensibly sent by teletype from Wright-Pat to Kirtland AFB.

William Moore, coauthor of the 1980 book *The Roswell Incident*, which began the revival of interest in the case, also claimed that Doty gave him secret information on UFOs in February 1981. Moore claimed to have received a briefing paper on "Project Aquarius," an effort to make contact with aliens, which was later mentioned on the television show. The paper mentioned other projects, including the one called Snowbird, which since 1972 had been testing a flying saucer "somewhere in Nevada," and it also referred to access restricted to "MJ-12." Thus an infamous code name was first introduced to the world.

In January 1982, Moore met Robert Pratt, a former *National Enquirer* reporter, and told him about Doty, his "Deep Throat" source. The two wrote a novelized version of the story, called *The Aquarius Document*, which was never published.

In 1984, TV producer Linda Moulton Howe, known for her documentaries on the cattle mutilations, claimed Doty had provided her with a look at "presidential briefing papers" about flying saucers. One paper described a meeting between earthlings and aliens who had landed at Holloman AFB, near Alamogordo, New Mexico, at six A.M. on April 25, 1964. There was a similar meeting at Edwards—reminiscent of Ike's trip there. Howe claims that Doty told her there were people who wanted the information to get out: It was time. He promised her that film footage of the meeting and other dramatic evidence of contact would be forthcoming. Actual images of aliens talking with earthlings! But the additional material never arrived; Doty told Howe that there were political problems, a change of heart.

Moore said that Doty provided him with more information from a whole aviary of bird code-named informants, and in return Moore promised to report back to Doty on the activities of UFO researchers. Moore, in turn, was collaborating on research with Stanton Friedman, the UFO researcher who had once worked for Aerojet and had a master's degree in nuclear physics. Moore also began to deal with Jaime Shandera, a film producer.

On December 11, 1984, shortly before he left his home in Burbank, California, for a lunch with Moore, Shandera was thumbing through *Variety* when he heard a rustling and then a thump as something dropped through the mail slot in his front door. It was a package wrapped in brown paper and postmarked Albuquerque, where Doty was stationed at Kirtland AFB. Inside was a roll of undeveloped black-and-white 35mm film. Moore and Shandera developed the film, and as the prints were drying they read the words "TOP SECRET/MAJIC/EYES ONLY" in the orange safety light of the darkroom.

It was the beginning of the Majic 12, Majestic 12, or MJ-12 story. The group had supposedly been formed by President Truman in September 1947 to investigate UFOs. Among the photographs was one of a document dated November 18, 1952—a briefing paper for President-elect Eisenhower stating that the remains of four alien bodies had been recovered two miles from the Roswell wreckage site.

The members of the MJ panel were a predictable but convincing group of high-level government and military officials and top scientists. If fictional, the list had been cleverly confected. It included Lloyd Berkner, a member of the CIA's Robertson panel looking into UFOs; James Forrestal, first secretary of defense in July 1947 and, famously, a suicide at Bethesda Hospital in May 1949; Gordon Gray, assistant secretary of the army, who later became head of the 5412 committee, Ike's inner circle for national security decisions; CIA director Walter Bedell Smith, who had discussed the psychological warfare implications of UFOs; MIT professor Jerome Hunsaker, the head of NACA, the predecessor of NASA; and Nathan Twining, commanding general of the Air Materiel Command at Wright Patterson Air Field, on record as believing that "the phenomenon is something real."

The group's alleged head—called "MJ-1"—was Vannevar Bush himself, who during World War II was head of the National Defense Research Council, in charge of the Manhattan Project, the Radiation Lab at MIT, and other important secret research programs

———

The MJ-12 documents were the biggest thing to hit the UFO world in years. Moore, Shandera, and Friedman did not make them public until 1985 because, they said later, they wanted first to verify their authenticity. But critics, led by longtime UFO debunker and *Aviation Week* editor Phil Klass, immediately attacked the documents.

All of a sudden the UFO world sounded as if it had turned into a covey of graphologists. The critique of the documents leveled the following charges: that the rubber stamp of the TOP SECRET banner was not made for the purpose but had changeable type (like old-style library due-date stamps) that bore a strong resemblance to one Bill Moore had used for his own return address; the presidential directive order number establishing the group was not consistent with the numbering system (secret directives, Friedman countered, would naturally not be numbered with the standard system); the documents bore no top-secret register number, and the classification "Top Secret Restricted Information" was not used until years later; the supposedly top-secret documents did not have the "Page ___ of ___ pages" indication that is standard; the document uses the form "Roswell Army Base," which was not used after 1943; the text includes the term "media," and uses "impacted" as a verb, years before such locutions became common; and, finally, the Truman signature on the September 24, 1947, memo seems to be identical to one on an October 1, 1947, letter to Vannevar Bush, right down to the distinctive skid mark on the H in "Harry."

A dizzying series of debunkings and counterdebunkings began, turning on watermarks and date formats, bureaucratic procedure and national security directive numerical formats. But the debates that burned up the newsgroups and UFO seminars seemed to lodge mostly on the dating matter: Phil Klass had charged that the MJ-12 documents' use of a zero before a single digit day of the month (as 01 August or 06 December) reflected a style that had come along only after the advent of computers. It was, he wryly added, also a format Moore had used since the fall of 1983 in his personal papers, as well as in "retyped" official documents he'd distributed. In response, Friedman pointed out several cases where the zero had been used, especially in military documents. The charge and countercharge, assertion and rebuttal, seemed to zoom in on that zero, circling the null, like a bug in a draining sink. It was more closely scrutinized than any circle or disc in the sky had ever been.

———

Moore revealed that in March 1985 he had received several anonymous postcards. They had been sent to his post office box in Dewey, Arizona, a previous address, then forwarded to his new home in Los Angeles. One card showed a photo, credited to the Ethiopian Tourist Commission, which pictured the African bush. The return address read, "Box 189, Addis Ababa, Ethiopia," but the card had been postmarked in New Zealand. On the back, a single-spaced, typed message read in part: "To win the war . . . Add zest to your trip to Washington / Try Reese's pieces; / For a stylish look / Try Suit Land."

For Moore and Shandera, "Reese's pieces" was a reference to the candy eaten by the alien in the film E.T. But they reminded Friedman of something else.

To validate the authenticity of the MJ-12 documents, Friedman had planned to visit the National Archives, where some promising military intelligence records had been declassified. The trio was particularly interested in Record Group 341. Friedman was scheduled for a lecture tour and asked Moore to go to Washington in his stead to check out the documents. Moore was not excited by the idea and told Friedman about the baffling postcards he'd just received. Then Friedman happened to mention that the records were not at the Archives themselves but in a branch in suburban Suitland, Maryland, and in the charge of a man named Edward Reese.

Moore and Friedman would discover a letter in the declassified records signed by Robert Cutler, Eisenhower's national security adviser, referring to a change in the time of an "MJ-12 meeting." Later, they realized that the box in which the memo had been found, number 189, was the same number as the post office box on the return address on the postcard.

Who had sent the cards? And if the letter had been planted in the archives, who had done so? Was it some kind of Deep Throat seeking to let the secret out, or a disinformer working from inside?

Robert Cutler's letter, found loose between two manila files, was dated July 15, 1954, and referred to "NSC/MJ-12 Special Studies Project." It was a carbon copy on onionskin paper bearing the red slash officially declassifying it. The letter notified Nathan Twining that "the President has decided that the MJ-12 SSP briefing should take place during the already scheduled White House meeting of July 16, rather than following it as previously intended." Friedman and Moore believed this innocuous, procedural note was the smoking gun that verified the existence of the MJ-12 group and the authenticity of the earlier MJ-12 documents, but critics attacked what quickly became known as "the Cutler-Twining memo." They noted that it did not bear the standard government eagle watermark, that Ike had no such meeting on July 16, 1954, and that Cutler was not even in the country on the day he was supposed to have signed the letter. (His aide James Lay signed a genuine memo on the same date.) Yet Richard Bissell, who had looked at top-secret documents throughout his career, concluded that he could find nothing obviously false about the letter.

The MJ-12 tales grew and acquired a backward history: Their origins were better known than their subsequent existence. In the accounts of people like conspiracist William Cooper, MJ-12 became linked to the Bilderberger clique and the Trilateral Commission (whose triangular symbol was said to have been derived from the markings found on a crashed flying saucer). In these accounts, MJ-12 killed JFK and forced Nixon from power because he wouldn't cooperate.

After the Lazar stories surfaced, an investigator named Robert Collins published an elaborate organizational chart for the MJ-12 organization, ostensibly dating from 1984–85, which included a prominent branch for operations at Area 51 and S-4. His source, he said, was William Moore. John Lear told in his lectures of a country club where the MJ-12 members held their meetings, playing golf and tennis in between considering the fate of the planet. According to these accounts, Henry Kissinger and Harold Brown had become members, along with Lew Allen of the Strategic Defense Initiative and Bobby Ray Inman, former deputy chief of the CIA and NSA boss.

To UFO and secret aircraft historian Curtis Peebles, the whole thing was very much in the classic American line of suspicious secret groups and "cabals." Twelve was a neat number; Peebles noted that one member, Harvard observatory director Donald Menzel, was a widely read UFO debunker and could figure as a Judas. He also referred to the twelve Jewish elders to whom the notoriously fraudulent Protocols of the Elders of Zion had been attributed. (Five of the original MJ-12 panel had "Jewish-sounding names.") George Van Tassel had been advised at Giant Rock by a "Council of Twelve" key followers.

The idea of a secret, conspiratorial organization charged with recovering and repairing saucers while hiding the existence of the UFOs had deep roots. Ufologist James Moseley wrote in Saucer News in 1956 of reverse-engineered saucers. "This type of saucer is not built by the American Government as we ordinarily understand the word 'Government.' As fantastic as this might sound . . . these saucers are actually built, operated and maintained by an organization which is entirely separate from the military and political branches of the Government that we know about. I shall call this secret project 'The Organization.' " Donald Keyhoe moved from suspicion of the Air Force in his first book to charges of outright conspiracy and cover-up in his next, The Flying Saucer Conspiracy.

But the MJ-12 project also sounded a lot like actual commissions formed to deal with vital questions of national security, such as the Killian Committee, which had recommended building the U-2, and the Gaither Commission, formed around the time of Sputnik. And Majic suggested the "Magic" code name of U.S. intercepts of Japanese codes. In fact, Larry Bland, editor of the George C. Marshall Papers, asserted that one of the Majestic-12 documents reproduced language from a 1944 letter Marshall had sent to presidential candidate Thomas Dewey regarding the "Magic" intercepts, with the dates and names altered and "Magic" changed to "Majic."

If they were fake, the MJ papers were well done. Were they too good to be true? Or was someone "inside" working to get the word out? Wasn't there too much in these documents? Reading them, one sometimes gets the sense that they're like a subplot in a badly written film trying to fill in background for the audience. It was also hard to imagine the social dynamic in the room when Eisenhower was told of all this, the Pentagon updating the president-elect on a day when he was scheduled for less than an hour of briefings—just forty-three minutes, one researcher concluded from a study of Ike's date book, for the entire world situation. Was he informed somewhere in those forty-three minutes that, by the way, we have found flying saucers from another system and have four alien bodies and a clandestine committee shrouded in secrecy deeper than that of the Manhattan Project?

Early in 1985, Richard Doty was transferred from Kirtland to West Germany. It was later charged that he had faked contact reports there and flunked a lie-detector test, both of which Doty denied. Then, in May 1988, he denied that he had shown Linda Howe any presidential briefing paper on Aquarius or anything else, or that he had even heard of MJ-12.

In 1989, speaking at the national MUFON convention, Moore confessed that he had been a double agent, spreading disinformation on behalf of AFOSI and spying on UFO believers in order to get closer to the truth and get behind the cover-up. Desperate not to be left out of the biggest story in history, he had collaborated.

Before long, Moore vanished from the UFO scene. Had he been telling the truth? Was this an AFOSI job? Was Moore really spreading disinformation? Or were his claims to have traded duplicity against his fellow youfers for access to hidden information themselves bogus? Was he a double agent or a triple one? Did he simply disinform Bennewitz, or did he, by himself or in collusion with Doty, fake the MJ-12 documents? Finally—given that his story seemed to reflect those of Moore and Doty—was Lazar's story a continuation of the same scheme, by way of Lear (who had "gotten interested" in UFOs in 1986 and cited the same April 25, 1964, date for the "meeting with aliens" at Holloman)?

To some of the youfers it hardly mattered whether the MJ-12 documents were real or phony. A number of apologists for the papers worked them into the trickle-out theory—they might not be genuine but they were preparation, a means of breaking the news gradually. Without being rank disinformation, they could still be part of the Big Plan from those On the Inside, preparing us with a softer version of the truth.

Linda Howe continued to stick to her story of seeing the briefing papers, but the record was not in her favor.

Was Doty the center of the disinformation scheme that involved Moore, perhaps John Lear, and eventually Lazar? Was he a loose cannon or a nut? Were the doubts about Doty's veracity a further muddying of the waters by the AFOSI or a recognition that he had gotten out of hand? Doty left the service and was hired as an investigator in the New Mexico Department of Public Safety. "I don't do interviews," he told me. "I can't talk about it. I'm still sworn to secrecy."

He didn't sound like a nut; he didn't sound like a hard-ass military type, either. He had a cowboy sort of twang and sounded quite likable. "I wouldn't ask you to violate your oath," I said.

I suggested that he might not have been treated fairly by the record of the whole matter.

"I don't worry about it. People can think what they want to think. The truth is known by those that matter."

"Did you distribute any false documents?" I asked.

He repeated that he could not talk about it.

"You feel you acted according to your duty?"

"Absolutely."

He suggested he might be able to call me back. I was amazed when he did. "We did not fabricate any of the papers," he declared. "I have been investigated and cleared."

Howe had brought the alleged briefing paper with her, Doty said. I didn't have to take his word for it. The meeting with Howe, he revealed, had been videotaped; the footage backed him up.

He denied having created any briefing paper, or having delivered it. He did not create the MJ-12 papers, he said, nor did he know who did. AFOSI thought that the culprit was most likely some private citizen. The implication was Moore and/or Shandera. Moore, he told me, was "a low-level source" for AFOSI. The FBI and AFOSI tried to determine the authenticity of the MJ papers but could not; they ran up against classification barriers. "A lot of time and money was spent trying to determine whether they were genuine," Doty said. "The FBI came up with fifty-fifty." He also denied being Falcon or Condor. Falcon was another man, he said, in his eighties if still alive.

I ventured some ideas: In the past, black programs had always been covered with stories. The U-2 cover was the weather-plane story. The Stealth fighter was covered in part by the phony A-7s with bogus electronic pods. Had AFOSI ever run deception programs using UFOs as a cover or diversionary story?

"Yes," he said, "you're right on the money. I've never worked on any, but there have been some. It's called 'legitimate lying.'"

Then I learned that Doty had been born in Roswell; his father had worked on the U-2 program.

16. The Real Men in Black?

Could Doty be believed? My instincts said so, but I could hear the doubtful saying, Well, what else would you expect from a disinformation agent?

Yet who was to say that the whole MJ project had not been some larger version of the sort of cover stories used to hide black programs like that of the F-117? What Doty had said gave me a new model: Dreamland not just as a reflecting funhouse mirror but as a malignly refracting crystal. To use the language of the low-observables engineers, this was not just passive stealth, bouncing inquiries off at oblique angles, but active stealth, generating false signals, "spoofing." For most of the Interceptors, the assumption has been that the military does not need to generate disinformation, that, as John Andrews once put it, the "natural mutations" of information would do the trick: The noise itself would be taken for signal.

But once you opened your mind to the possibility of active disinformation, all kinds of questions came up. Who, some UFO buffs could ask, would most likely have the resources to fake the celebrated Santilli film of the "alien autopsy" claimed to date from the Roswell crash? If such official duplicity was common, what, then, could be trusted? For years, youfers had been charging the government with a cover-up, but now, ironically, the opposite had to be considered: The government may be manipulating, even generating, UFO reports to conceal secret programs.

This possibility was made more intriguing by a report that appeared in the summer of 1997 suggesting the CIA and Air Force had been happy to have many sightings of the U-2 and the A-12 classified as UFO sightings simply to hide the existence of the planes. This was natural mutation at work. The report came in an article by Gerald Haines, the historian of the National Reconnaissance Office, which appeared in the CIA publication *Studies in Intelligence*. In "A Die Hard Issue: CIA's Role in the Study of UFOs, 1947–90," Haines declared that "over half of all UFO reports from the late 1950s through the 1960s were accounted for by manned reconnaissance flights (namely the U-2) over the United States." This was especially true during the early days of its development, when the aircraft, not yet painted black, were silver.

Such confusion was apparently officially encouraged, but there was no evidence that it was actually suggested or directed. Later, things may have become different. Haines recorded that one branch of the CIA, the OSI's Life Science Division, had "counterintelligence concerns" in the seventies and eighties "that the Soviets and the KGB were using U.S. citizens and UFO groups to obtain information on sensitive U.S. weapons development programs such as the Stealth aircraft." Such information mutations led to active disinformation. The Air Force, Haines reported, was forced "to make misleading and deceptive statements to the public in order to allay public fears and to protect an extraordinarily sensitive national security project."

To understand how this might have evolved, one has to understand the mind-set of the Reagan era, a period I came to think of as the Second Cold War, a renewal of hostility and fear after the détente of the seventies. Its beginnings can be traced to Ronald Reagan's characterization of the Soviet Union as "the Evil Empire"—the resemblance of the era's language to that of the film *Star Wars* is significant. The ideology of this second war, however, can be traced to a group of advisers who believed that the Soviet threat was underestimated because it had been cleverly disguised.

I had met one of the men who believed this. He had been a deputy director of the CIA in the early seventies and was now in private business. He introduced me to the concepts of *maskirovka* and *dezinformatsiia*— disinformation. *Disinformation* was the buzzword of the resurgent youfer conspiracy theories of the eighties.

We met at a Denny's restaurant near Washington. Somehow we got to talking about the Blackbirds. He shook his head in lingering wonder at the Skunk Works, and recounted the speed with which they had built the Blackbird. He seemed nostalgic for those days, the years of the Evil Empire and Star Wars, when "active counterintelligence"—Doty's "legitimate lying"—was in vogue.

The man explained that he had debriefed a Warsaw Bloc defector, a Czech named Jan Sejna, who told him that practically the whole Czech air defense system was phony: wooden missiles, fake troops. The Soviets, he said, would run whole schemes of bogus radio traffic and move out dummy equipment during the periods when U.S. satellites passed overhead. It sounded all of a piece with the infamous Soviet Air Force Day fly-by—the Hollywood trick of running the same bombers past the reviewing stand again and again—that had inspired the bomber gap.

This, he explained, was a technique long beloved of the Russians: *maskirovka*—the use of disguises, tactical and strategic, of fake SAM sites, of airplanes that flew around in circles in May Day–type parades, "bluffing up" the bomber forces to suggest a bomber gap that the U-2 would have to seek out and debunk. There was a Russian tradition, apparently, of the military Potemkin village.

There was also *dezinformatsiia*, he told me, the creation of phony programs, documents, and informers. And who was to say the Soviets weren't doing this on the strategic as well as the tactical scale? Such thoughts were freely accepted at hard-right institutions like the Hoover Institute. For men such as Sam Cohen, "the father of the neutron bomb," this raised larger questions: Were the missiles the Soviets seemed to be placing in silos really there, or were they empty containers? Were the real missiles stashed away on railcars traveling the vast and trackless central expanse of the Soviet Union, in violation of all SALT treaties? This was a popular line of thinking in the Reagan years; later the proponents of the B-2 Stealth bomber suggested that the huge flying wing would roam the countryside searching out those missiles. In the traditional Cold War mode called "mirror-imaging"—if they've got a weapon, we also must have that same weapon—it could also be taken as justification for an American effort in *maskirovka* and disinformation. Thus the eighties became a period of "proactive" deception—American disinformation to cover secret programs.

I had doubts about the whole premise. What if the defector himself was disinformation, a plant? What if the fake missiles were real? If the goal had been to spread doubt and uncertainty, it had worked. We, the enemy, were frozen if we bought the story—either way. In this light it was worth considering whether MJ-12 or at least the Bennewitz disinformation effort had been a kind of counterintelligence maneuver.

The answer would likely never be known. John Pike, the Federation of American Scientists expert on secrecy, was convinced there had been many active counterintelligence programs surrounding Star Wars. He met a man who had presented himself as a journalist for an industry newsletter and who claimed to have a lot of useful information about Area 51. But, Pike would recall, "he seemed more interested in telling than in asking," and he subsequently seemed to vanish into thin air. Active disinformation, he felt sure, had helped cover Star Wars programs. And Star Wars' relation to Dreamland was another mystery.

On March 23, 1983, Ronald Reagan delivered his famous "Star Wars," or "Strategic Defense Initiative," speech. Within a year a huge effort was under way to develop all sorts of high-tech antimissile weapons and energy beams. Black budgets for weapons increased from $892 million in 1981 to $8.6 billion for fiscal year 1987—and Dreamland would have been the natural place for testing the aircraft, UAVs, lasers, high-frequency radio weapons, and other technologies. The Ballistic Missile Defense Organization that had grown from the Star Wars initiative was also the likely developer of the Tier III UAV, the legendary bat plane that had been spotted around Dreamland.

These years naturally saw increased efforts in security as well, especially in the area of disinformation, based on a reading of Soviet disinformation and *maskirovka*. And psychological investigations of various types—such as those developed at the R&D fringe of the CIA back in the days of blind LSD experiments and brainwashing—seem to have increased.

The Lore is full of references to mind control experiments, but holographic images are also a real possibility: The test of the ability of equipment to create illusions would have been just the sort of thing that could produce strangely moving lights above the Jumbled Hills.

Reagan believed in *Wunderwaffe*, wonder weapons. This belief was deeply rooted in his Hollywood past, as the historian Garry Wills showed in his book *Reagan's America: Innocents at Home*. The Star Wars–era ray guns were anticipated in Reagan's Brass Bancroft films of the thirties and forties, which introduced similar themes. *Murder in the Air*, for instance, described an "inertia projector," a kind of force beam. Reagan was idealistic about the lasers: They were defensive, just as the Lone Ranger uses his silver bullets not to kill people but to knock the guns out of the hands of the bad guys. It was a dream of peace without bloodshed.

With the military buildup, maneuvers and exercises increased. NATO was flying entire armored divisions across the Atlantic to show that it could respond to a Soviet invasion. Counterintelligence didn't want to be left out. How do you "go on maneuvers" in counterintelligence? Well, you run an active disinformation campaign, targeted at producing a widespread belief in something. The target could have been the American public at large or simply the UFO community, just as the targets for SAC or for Stealth were, in practice runs, the cities and towns of America.

The counterintelligence professionals included the Air Force Office of Special Investigations. AFOSI, in charge of such matters as theft, drugs, procurement violations, and the personal security of officers, is also in charge of counterintelligence, which means preventing spying or other information leaks. Some counterintelligence is defensive but most is offensive or proactive—the creation of covers, such as the deceptive A-7 aircraft that were displayed as cover for the Stealth fighter.

Some active disinformation documents, including faked Air Force letters and reports, have been convincingly traced to AFOSI and to a Colonel Hennessey. He was part of the AFOSI's PJ section—which handled counterintelligence for secret programs. When Congressman Steven Schiff of New Mexico wrote the General Accounting Office inquiring about any government records having to do with Majestic 12, the only interesting item the GAO came up with was a message dated November 17, 1980, from the operations division of AFOSI, containing the phrase "MJ Twelve." AFOSI concluded that the message was a forgery.

Whether they were real, active disinformation, or amateur fakes, one of the strongest pieces of evidence against the MJ-12 papers is the reference in one of them to Area 51. By the best evidence, that designation had not been bestowed by the Atomic Energy Commission until 1958. (Peter Merlin tracked it to the test site bulletin, and mundane changes in phone numbers.) The area had not been used for anything to speak of until Tony LeVier flew over it in 1955.

However, there is incontrovertible proof of the existence of "alien" craft at Area 51—Soviet ones. The military tried to hide their existence, but it was secretly very proud of the captured Soviet airplanes, furtively brought into the area. And since they were there—denied, hidden—who was to say that yet other alien craft were not there also?

17. Red Square, Red Hats, and STUDs

PRESS BUTTON FOR PLEASURE, reads the neatly routed Formica sign on the chain-link at the Shamrock Lounge, near Lathrop Wells, west of Dreamland. The Shamrock is one of several legal cathouses in Nevada's Amargosa Valley, most little more than a collection of trailers, linked together and fenced off.

If you had been inside the Shamrock, or in the café at Lathrop Wells, where old men linger, sipping coffee and sopping up their gravy with sourdough biscuits, late on the morning of March 26, 1984, you would have been distracted by the boom of an airplane hitting the ground.

It was the sound of Maj. Gen. Robert Bond plunging into the ground and to his death after losing control of his aircraft and smashing into a mountainside.

Secret planes become unsecret when they fall out of the sky. The news of a mysterious airplane crash near Dreamland and the fact that a general was in the cockpit meant the story could not be contained. There was immediate press speculation that Bond had been flying a "supersecret new stealth airplane." Among UFO watchers, the speculation went further—had Bond been testing one of the recovered saucers? As recently as 1991, in his book *Cosmic Top Secret*, William Hamilton declared, "The Air Force refused to say what type of plane Bond was flying at Area 51, but it seems highly irregular for the Air Force to use a three-star general as a test pilot. Is it possible that the general was test flying a recovered alien spacecraft?" In fact, Bond had been flying a MiG-23.

Bobby Bond was known around the Skunk Works as a stickler and worrier. A hard-driving Tactical Air Command fighter type, he was also cleared on Stealth and other secret programs. Bond's crash brought the foreign technology program to light. It was run by what was at one time called the Foreign Technology Division out of Wright Patterson—exactly the place where the saucers were supposed to be hidden. Foreign tech indeed.

The Lore held that other alien technology was buried more deeply in the legendary Hangar 18, and alien bodies—some said dead, some said living—were kept in tanks of liquid and cryogenic coolers. Where but the Foreign Technology Division, the believers ask, would the Roswell bodies have been stored?

————

The only piece of alien flying technology ever photographed and positively identified at Dreamland was not from Zeta Reticuli or any other system. It was a MiG-21, captured in John Lear's shot from the lake edge in September 1978. That image confirmed what had long been suspected: that a program existed for testing aircraft captured, stolen, bribed, or otherwise purloined from the Soviet bloc.

Like a real-world shadow of the UFO testing programs of the Lore, the "Red Hat" squadron program was highly secret, in order not to compromise the sources of the planes—and the spare parts, engines, and tires needed to keep them flying. The important secrets had to do not with enemies but with allies, as Frank Powers was given to understand: The U-2 flights to be protected at all costs were not those over the Soviet Union but those over Israel and Egypt, aimed at the waning power of Great Britain and France.

It began in 1953, when the Air Force's Foreign Technology Division (now Foreign Aerospace Technology Center) at Wright Patterson Air Force Base first flew a Yak-23, smuggling it out of Eastern Europe, testing it from the Dayton base as a U.S. "X-5" painted up in American insignia. When the testing was finished, the airplane was smuggled back inside the Iron Curtain.

The "Black Yak" was followed only years later by a series of tests of MiGs, called by the code names "Have Drill" and "Have Doughnut." This was the work of the 4477th Test and Evaluation Squadron, the Red Hats, who wore red stars on their patches, and the Air Technology intelligence center.

When Kelly Johnson had pushed for the Blackbird as a high-altitude interceptor, he was thinking of war against the Soviet Union, of fleets of incoming bombers, to be dispatched by look-down shoot-down radar, and missiles. The Air Force envisioned fighter combat in largely the same terms: The fighter bosses believed combat would take place without the two opposing fighters ever seeing each other. They would lock on by radar at long distance. No guns were necessary.

But dogfighting returned in Vietnam, with U.S. fighters facing North Vietnamese MiGs. The F-4 Phantom was losing fights with MiGs at a disturbing rate—one Phantom downed for every two MiGs.

In the Six-Day War, the Israelis acquired a number of the Soviet planes from captured airfields, defectors, or—in one case—when Libyan pilots landed at a Sinai base they did not know had already been taken by the Israelis. They are the probable source of a MiG-17 obtained by the Defense Intelligence Agency in 1967, the first one to fly in Dreamland. Later, the United States would acquire—just how remains a mystery—a MiG-21, Su-22s, and MiG-23s, and by the nineties even an Su-27 Flanker, the most advanced Russian aircraft.

After France cut off sales of Mirage fighters to Israel, to placate the Arab nations that provided its oil, the United States struck an agreement to sell the Israelis planes and made captured MiGs part of the price.

Soon men in the ranks at Nellis AFB began jokingly referring to the box of restricted airspace around the secret base as Red Square.

In the project called Have Drill, the Red Hats flew the MiG-17 from Groom Lake in simulated dogfights over the desert to figure out its strengths and weaknesses. The MiG performed better than the Phantom at low speeds—it could turn "inside" the F-4 every time—but if the F-4 pilot kept speeds up and stayed outside and behind, he could win. Using the tactics developed at Groom, the Navy produced a film called *Throw a Nickel on the Grass* (a line from an old Navy fliers' song) and brought in classes of pilots for retraining. By the end of the war, the kill ratio had shifted to eight-to-one.

At the 1969 Tailhook convention in Las Vegas, the talk among the pilots turned to MiGs. A number of admirals were flown up to Groom and put through the paces in the captured aircraft.

In the eighties, with more aircraft acquired through Afghanistan, a new program called Constant Peg was set up at Tonopah. The Red Eagles, as they were called, were part of the same 4477th Test and Evaluation Squadron. They trained Navy as well as Air Force pilots and, after carrier-based fighters shot down two Libyan MiGs in 1989, a Pentagon spokesman, in a moment of pride overcoming tact, bragged that the successful pilots had been trained in enemy aircraft.

With the collapse of the Warsaw Pact and the Soviet Union, the foreign-technology boys set their sights even higher. It was the military yard sale of all time, and by 1993 the American taxpayer was spending half a billion dollars per year for "foreign material acquisitions."

Trader (aka Paul McGinnis) tracked down the cost of running one evaluation program—he could rattle off the number by heart—the program with element number 207248F. The program behind the number was called STUDs, for "special tactical unit detachments." It is hard to believe that any connotations with regard to this acronym are other than intentional. From fiscal 1993 to 1994 STUDs went from $885,000 to $20 million, and to $118 million for 1995.

Many of the foreign aircraft have probably come into the country by surreptitious sale or bribe. They may even include advanced prototypes purchased from a renegade general or engineer. But there was a limit to the process—a classic catch-22. All military systems are supposed to involve fair competition among different contractors or suppliers, with a request for proposals and evaluation. But since the "source" of the MiGs was not only "sole" but secret and clandestine, the Pentagon could hardly hold competitions among the corrupt Warsaw bloc colonels or Third World defectors who could provide the hardware.

MiGs might be confused with saucers, but they still looked like fighters when you got a clear look. They did not look like the strange flying beast I had seen at the boneyard, nor did they look anything like the far stranger shapes that had flown above Groom Lake and whose cousins were about to come into plain sight.

18. El Mirage and Darkstar

TUMBLEWEEDS CLEARED, read the sign by the side of the road, a fair indication of the nature of local enterprise. This was the desert east of Palmdale, California, center of America's high-tech aerospace industry.

I had passed Plant 42, where the B-2 bomber was hatched. The new Skunk Works was nearby, its huge hangars crisply painted in a gray worthy of the most shipshape vessel in the Navy.

The road ran east from "Aerospace Valley," formally known as the Antelope Valley, although no such animal has been seen there for years. Instead, you saw the new malls, Kmarts, and fast-food franchises, giving way eventually to acres of sod farms and fruit trees. The map showed a huge expanse of lettered avenues and numbered streets, the projection of a vast city dreamed up by some wildly optimistic boosters. It centered around a planned super-airport that had never been built; the map even showed the runways and terminal sites.

It resembled one of the Nazca earth markers that Erich von Däniken thought represented a landing strip for the gods. It was a dream airport, as if for the ghost craft watchers frequently saw here—the mother ship, the giant triangles, the bats and whales. An airport of the imagination, I thought, like the airbase of the imagination at Dreamland.

———

Out of the corner of my eye I caught the same evil-looking shape I had seen at the boneyard many years before. The spot was Blackbird Park, a strip of grass near the Lockheed Skunk Works where examples of its proudest works are parked: A-12 and SR-71 Blackbird aircraft. Beside them sat the little craft I now knew was a D-21 drone, the last of the Blackbirds, kept secret for years, which had flown from the back of an SR-71, then from beneath the wing of a B-52.

It was a transitional design between a manned spy plane and a UAV—unmanned aerial vehicle. The engineers called it a "parasite." Once released from the back of the Blackbird, the D-21 was automatically guided. Out of control of any ground station, it would pass over a target—the denied area—and photograph it, then fly to a friendly country and land by parachute.

It weighed several tons, a chunky cylindrical shape with stubby wings and tail that clearly suggested the SR-71; it was like a larval version of the big craft. The whole project was called Tagboard.

The D-21 foreshadowed a new generation of UAVs. I wondered if many of the shapes flying out of Dreamland were not of this type. Unmanned craft could and did take on shapes that were, in the words of many observers of things flying above Groom Lake, "otherworldly." Because they did not have cockpits or windows, because they did not need to provide protection for a human pilot, they could be more batlike or more saucer-shaped. They could be pumpkin-seed-shaped.

How many UAVs, flying secretly, had been taken for UFOs? And how many others had been hidden as well as the D-21 had been? How many aerial sharks and mantas?

Mine was not the only suspicion. The first assumption, incidentally, of those who encountered flying saucers was oftentimes the same as my own: They must be secret planes of some sort. The famous front page of the Roswell Daily Record, now reproduced on thousands of T-shirts and in many books that proclaimed the "capture" of a saucer or disc, also included a story on "man in the street" reactions. H. M. Dow of Roswell declared, "I have come to the conclusion that there are some disks flying and I think it is an experiment of some tactical branch of our armed forces." One Rolla Hinkle opined that "the United States government is trying out something new. These disks may be radio-controlled instruments of some kind."

———

I drove past old ranches, with corrals jury-rigged from wire and discarded doors, looking for a very different kind of airplane. I was heading for El Mirage. The dry lake there had long been a favored spot for hot-rodders and motorcyclists, who cut loops and doughnuts into its surface. For artists, too, it was a useful canvas. In the late sixties and early seventies, earth artists had created temporary sculptures here. Inspired in part by the vast canvas of the desert, one had poured strips of asphalt on El Mirage in an X shape that looked from the air like the little x of the airstrips at Dreamland when Kelly Johnson and Tony LeVier had first flown over them. Another artist had sliced long trenches into the lake bed to define "negative space" and what were called "nonsites." Weren't the Air Force and CIA into "nonsites" too when they ran bases they wouldn't acknowledge?

I had just parked next to an old aircraft boneyard when I caught sight of it: a tiny fleck that came closer, turning into what looked like a giant white paper plane, with wing tabs turned down, wheeling over the small airfield. But the strangest thing was its nose—it had no cockpit, no windows. It looked blind.

There were no windows because there was no pilot. This was Predator, the most recent UAV, flying for the CIA and the military, being tested at the El Mirage desert airstrip of its builder, General Atomics, Inc.

———

For a long time, we didn't even have a good name for these things. Once, they were dismissively called "drones," then "remotely piloted vehicles." By the mid-nineties, the term of choice had become UAVs, and in the Pentagon the field was chic. A new generation of UAVs was arriving, relying on advances in electronics and computing, miniaturized sensors and cameras and relay systems. Today's UAVs are spy planes; tomorrow's will be fighters.

For years, UAVs inhabited a world of their own, a shadow of a shadow. Overlooked, ignored, they never attracted the kind of attention the black planes did. Before they became fashionable, how long had they been flying out of Dreamland?

Even the most famous of imagined Dreamland projects may have been a UAV: The "Glossary of Aerospace Terms and Abbreviations" in the September 1994 issue of *Air International* claimed Aurora is an acronym for AUtomatic Retrieval Of Remotely-piloted Aircraft. And the builder of the huge Perseus UAV for NASA was a company called . . . Aurora.

Hovering high above unfriendly countries, their proponents say, UAVs can relay via satellite to distant ground stations video, radar, or infrared images of anything that moves. "Lingering" is the favored term. The Predator, for instance, can fly three hundred miles and "linger" for up to two days in the air, where it is virtually invisible to the human eye and difficult for radar to spot. (Despite a fifty-foot wingspan, it shows up only as a square meter radar "signature.")

Proponents have proclaimed the dawn of a new era in aviation and a new kind of pilot—the right stuff of the future. The joystick in the cockpit may be replaced by one on the desktop, and Top Gun may be replaced by Captain Nolo—traditional Air Force lingo for "no live operator."

Like robots of any sort, UAVs have the advantage of requiring no room and board, no training or food. They can pull more G's than human pilots (fighter aircraft are limited in their acceleration and deceleration not by the strength of their airframes but by the G-tolerance of the human body). Cases of "temporary interruption of consciousness"—blackouts—have been suspected in several crashes over the last few years, including General Bond's. UAVs cannot be held hostage or suffer torture. Politically, UAVs benefit from the new post–Cold War/post–Gulf War emphasis on inexpensive high-tech weapons that avoid putting human lives at risk. And of course UAVs cost less than manned aircraft. Predator was tagged at just $1.6 million per craft.

A couple of days after I saw it in the air, four Predators were on their way to the former Yugoslavia to conduct round-the-clock observation of forces on the ground. Predator, a so-called Tier II UAV, follows the Tier I "Gnat 750," which was less successful when tried out by the CIA from Albanian bases.

————

That afternoon I drove back west, straight up to the gray and blue buildings of the Lockheed Skunk Works—it was the post–Cold War Skunk Works now, as neatly groomed and carefully patrolled as any Hollywood set, properly outfitted as the high holy of American aviation. I had come to watch the unveiling of Darkstar.

Inside the hangar called Building 602, we were given press kits in neat black folders. Representatives were there to brief us. Until 2:28, when the curtain was to be pulled back, "the configuration was sight sensitive" and therefore officially classified. Lockheed, Boeing, DARPA, and DARO, the Defense Airborne Reconnaissance Office, were all partners in the project, and the craft, they told us, would go from drawing board to first flight in an unprecedentedly short twelve months.

"How," I asked later, "was such rapid development possible? Were there other programs that helped?"

They were, the DARPA man said, able to rely on experience from other programs.

"Could you tell what those programs were?"

"I could," he said, "but I won't." General laughter ensued.

The room was darkened. There was a great rumbling sound from above. I looked up and saw that the yellow roof crane that spanned the whole hangar was sliding slowly in the dark, a cluster of orange lights on its center, pulling back the black curtain. As stirring music played, dry ice spread a soft and ghostly fog around the craft: a white object that looked like nothing so much as a flying saucer with a large porthole. It took a few seconds to see that narrow wings grew from the saucer. It was just like the rollouts I had seen in Detroit for new cars—music, stage effects, lights: technology as theater.

One eager young PR person running around seemed to have nothing to do. I collared him and asked what music was playing. He disappeared and returned in a few minutes with the answer: It was from the Disney film *The Rocketeer*, based on a comic book about a man with a rocket back-pack. The film and the music evoked the romantic days of aviation, when Howard Hughes was setting flight records and making movies, and the alliance of Hollywood and aerospace was being formed. The name Darkstar was taken from John Carpenter's mid-seventies film about the crew of a roving spaceship.

Darkstar was the new so-called Tier III Minus UAV. The "Tier" desig-nations are DARPA project names, bestowed by a law, the "Section 845— Other Agreements Authority," that gave DARPA special powers for prototype development outside the normal channels of Pentagon pro-curement procedures. The Tier designations had been dreamed up to de-lineate the pecking order of UAVs by size, cost, and stealthiness. No one knew how, but Lockheed had managed to develop Darkstar in a matter of months. Rumor held that Darkstar was the son of a UAV called Tier III. By interesting coincidence, that name echoed "TR3A," the name of the Manta, the craft Steve believed he had spotted in Roswell. Tier 3—TR3. Was there a secret meaning? Or just a general confusion?

Predator is Tier I; Tier II Plus or "Global Hawk" was being con-structed by Teledyne Ryan in San Diego and could fly at 65,000 feet for twenty-four hours or more. Darkstar was Tier III Minus. It cruised at 180 miles per hour using a single jet engine buried inside what looks like a porthole. It could fly as high as 45,000 feet and survey some 1,600 square miles with synthetic aperture radar or electro-optical cameras. But its flying-saucer-like shape would make it more stealthy than the Tier II Plus. It had been, the briefers said, "optimized for low observables"; in other words, made to look like a saucer to avoid radar detection.

———

By talking to Interceptors and their network, I got an idea what the programs that could have aided in Darkstar's creation might be. One program was the Senior Prom stealthy cruise missile that had been tested at Dreamland. Another was Tier III itself, which the Lore said was also called "Q." According to various accounts, it was a successor to the Aurora debacle, an offspring of Lockheed's unsuccessful, alternative design for what became the B-2 bomber. It was a flying wing with a 150- or 220-foot wingspan. Others said that it, too, was a debacle. Two had been constructed and flown, manned, from the Groom Lake runway, but the program had been canceled because the cost of the individual aircraft had risen to nearly a billion dollars.

———

After the smoke and the oohs and aahs faded away, I talked to Maj. Gen. Ken Israel, the head of DARO, which along with DARPA, the agency that gave us the original Stealth fighter, had developed Darkstar.

General Israel used to fly in an EB-66 electronic spy plane probing Soviet electronics defenses. Now he quotes Shakespeare and touts the future of UAVs as a revolution in aviation. Israel's leading arguments for UAVs are humanistic: "In the next century, we will definitely rely more on pilotless aircraft to place people out of harm's way." But he also speaks in the terms of the new Pentagon fashion—"infowar." "We need to know what's on the battlefield before we get on the battlefield." With its ability to linger over an area, Israel says, a UAV can "view the battlefield with impunity." It can give the generals not just desktop infowar but *real time* infowar.

Look, too, he says, at "cost of ownership." The SR-71 Blackbird costs $38,000 an hour to fly, and a U-2 $6,000 an hour; a UAV costs only $2,000 an hour. These were craft for the post–Cold War world: cost-conscious, self-promoting, and aimed at very different enemies than Curtis LeMay's bombers had been.

The logic for UAVs had been obvious to some for years. Kelly Johnson predicted twenty years ago that they were the future of military aviation.

"UAVs are part of the great American tradition of substituting technology for human beings," says Randy Harrison, a member of the Darkstar team at Boeing. The Gulf War, and especially the difficulty of locating SCUDs on the ground, gave impetus to UAV proponents.

While for most American TV viewers the Gulf War seemed a model of information efficiency and intelligence gathering, Gen. Norman Schwarzkopf and other generals complained about their lack of "real time" information. The images we saw of smart bombs riding lasers down air vents were actually films, carried back to base and developed. Real information about enemy targets was much harder for the generals to get from space or the air. By the time satellite and other images reached the field from Washington, the tanks had often moved, the SCUDs shifted.

To be sure, this was the classic case of fighting the last war, but it also offered a look at the information war of the future.

General Israel and his friends imagine a war fought with batlike robot planes. Their strategy for overcoming conservative resistance is clever: "We are like Billy Mitchell," he said, invoking the prophet of airpower and his struggle for acceptance. The aspirations of UAVs to be real fighting aircraft are hinted at by their names: Hunter, Raptor, Talon, and Predator—pretty aggressive for mere reconnaissance craft. There is no reason at all, Israel says, that UAVs could not take over the job of the manned interceptor—that Captain Nolo could not supplant Chuck Yeager. And if the Pentagon goes to war with UAVs, won't the TV networks need them too? They will act as the high-tech equivalent of the news chopper.

The next step will be to use UAVs as target designators: eyes in the sky that will "paint" targets with lasers for smart bombs to ride down. The incentive for the UAV to replace the fighter, despite our affection for the chivalry and heroism of the dogfight, also comes as a result of the Gulf War. The Vietnam syndrome has been replaced by the Gulf War syndrome: total intolerance of casualties or the national humiliation of having pilots become prisoners displayed for the TV cameras.

Cases in point go back as far as Francis Gary Powers in 1960. But another conveniently popped up the very day after the Darkstar unveiling, when an F-16 was shot down over Bosnia carrying pilot Scott McGrady. With UAVs, there would be less need to send manned aircraft over such areas, and considerably less chance of pilots becoming hostages or pawns. A couple of days after McGrady was rescued, the decision was made to send the Predator over Bosnia. Had it been used earlier, it might have warned of the SAMs on the ground.

―――――

Studying Darkstar at the unveiling was a man in a blue fatigue cap and a leather A-2 jacket of the sort pilots wear. Lt. Col. Jim Greenwood was an RSO—the observer or backseat man in an SR-71—from 1986 to 1990. Now that America's dearth of aerial reconnaissance tools had led the Pentagon to pull the Blackbird out of mothballs, he was getting ready to fly again.

In the meantime he had become a proponent of UAVs—one of the few within the Air Combat Command, the fighter pilot's command. "Hey, it's a pilot's air force," Colonel Greenwood admits. Some pilots will resist UAVs to their last breath. But as for computers replacing the "human element" at the controls, Greenwood notes, that began happening long ago.

Computers fly airplanes much more often than pilots like to admit. Commercial airliners full of passengers are more readily trusted to computer systems than to human pilots during bad weather landings. Many aircraft, such as the F-117 Stealth fighter, are unstable without controlling computers.

The first controllers for Darkstar, Greenwood told me, would be trained pilots. "But in the future, you might take people straight off the street and give them pilot training, instrument rating, and then have them stop flying real planes and go to UAV school." The prospect of video-game stars taking over for the Top Gun hotshots did not seem to faze the colonel. "Gotta go," he said. "I've got a date with a T-38." Not half an hour later, he was arcing skyward at a steep angle, in the sort of airplane that may one day seem as quaint as a Sopwith Camel.

―――――

No one was quite sure yet whether the operator of a UAV should still be called a pilot. Captain Nolo flew the drones of the past, but today's UAVs don't necessarily need any pilot at all. Darkstar is programmed to roll out of the hangar, take off, fly its mission, land, and return to the hangar without human intervention. The Predator, by contrast, is directed by a joystick kind of mechanism. Darkstar uses Global Positioning System satellites to determine its location. Its flight plan can be changed in mid-course, but its interface is a series of maps and graphs of way points, a software system manipulated by mouse and keyboard.

————

The first official squadron of Air Force UAVs, I learned, the Eleventh Reconnaissance Squadron, commanded by Col. Steven L. Hampton, was already taking shape at Indian Springs. This, by happy chance, was also where Bob Lazar was debriefed. It was one of the places where legend said saucer wreckage was stored.

How many UFOs were UAVs? I still couldn't say, nor could anyone. But the Air Force was about to give confirmation of the existence of one of the strangest shapes that had ever been spotted in Dreamland.

19. The Remote Location

"Long ago," the general's speech began, "in a galaxy far, far away" . . . It was a reference to *Star Wars* that may or may not have been freighted with implicit criticism of the $40 billion weapons program of the same name. He was introducing a plane that the Interceptors had seen and talked about for years, though the military had denied its existence. A black plane that had flown only at Dreamland was coming into the light.

Outside the Air Force Museum in Dayton, a color guard presented arms; dignitaries straightened their ties. On the back of the chair in front of me was stenciled a collection of numbers and letters and the words USAF CHAIR FOLDING.

Interceptors were there in number. A man with a name I knew only from the Internet, where he provided specifications of aircraft present and past, right down to the tail numbers, wore a Lockheed Skunk Works T-shirt—a bit tactless, I thought, considering this was a Northrop project. He took pictures of everything that moved, including one of the cargo planes that flew overhead, then looked around a little sheepishly.

Before coming to the Air Force Museum, this airplane had been legendary at Groom Lake, open only to the handful of officers with all the necessary clearances. Again the strange principle seemed to hold true: To get close to Dreamland, you had to go far away. In Ohio we were getting a glimpse into the heart of Nevada, at a plane that had somehow traveled from rumor and suspicion to commemoration with no visible stops in between. Perhaps this was why the ceremony in Dayton seemed an odd mix—half confession, half celebration.

When the curtain was drawn, it was clear this was the ugliest aircraft most of the audience had ever seen. It was long and stubby-winged, hard to put the shapes together in your head to make a whole you could imagine flying. It looked like someone's effort to build a big fiberglass boat from magazine plans, abandoned halfway through. From the rear, it looked like some sort of modern architectural model, a building in Brasília, say, the exhaust vent a rising curve, like a concrete amphitheater.

It actually looked like Shamu, the star whale of Ocean World, and this became the nickname that stuck. The men who built it called themselves "whalers" and wore little lapel pins in the shape of a whale. That way, they could go out into the soft liberal world full of save-the-whale types and blend right in.

They had worked in a hangar beside the Stealth prototype Have Blue. For a long time, each of the two groups was forced to stay inside while the other was outside with its airplane: special access, need-to-know. It was the other team, getting a glimpse, that called it the Whale.

Whale was right: For the Interceptors, this was a kind of Moby-Dick, a great white whale of a genuinely mysterious flying object, long-sought, long-denied, legendary and mythical, now finally admitted. But material replaced mystery with a thud: The physical object was rough and ugly. It was also weird. The first thing I thought after seeing it was Who, describing such a thing floating over his head, would have been believed? Who then could not have had his certainty shaken that all flying objects were of terrestrial origin? With revelations surprising as this, how certain were we that there might not indeed be a Hangar 18, here at Wright-Pat or elsewhere, where alien bodies, creatures, parts, wreckage might be hidden?[1]

"There was a remarkable esprit" to the project, the speeches all agreed, born of the isolation of "the remote location," the silence of the black world, the camaraderie of the initiate. "We even did our phenomenology work in remote locations," said Stephen Smith, one of the top managers for Shamu. By phenomenology he meant radar testing. You could see the huge antenna from Freedom Ridge, and the big balloon balls used to calibrate it. These words suggested to me, however, the same old problem with Dreamland: that of knowing what was real and what was mere perception, speculation, rumor, fantasy. Of seeing and believing.

Patriotism, the dignitaries claimed, drove the project, even though it was not wartime. Steve Smith recalled "a strong sense of patriotic urgency with respect to Warsaw Pact nations at that time." In other words, fear of a big offensive in Europe that would overwhelm Allied ground forces and cause the United States to go nuclear. Shamu was designed to hover above a battlefield, using radars to direct thousands of "precision-guided munitions" at the hordes of invading tanks.

———

"There's a reception inside, under the B-36," the museum director announced, and the Whalers headed inside to stand under the wing of the huge plane. According to a sign, this was the last B-36 ever to fly, when it was ferried on April 30, 1959, from the boneyard at Davis-Monthan to Dayton.

At the reception, Steve Smith explained that he was working in Iran in the seventies, helping the shah's air force with its new F-20s, when the call came from his boss. He was being sent back for a special project, but was to be told nothing else until he was "brought in."

He was briefed on the third floor of a dark parking garage at a hotel in the San Fernando Valley, he recalled, "like Deep Throat. It was real cloak-and-dagger stuff."

The man in charge of the Tacit Blue program also stood beside the B-36. Jack Twigg was an Air Force colonel, detached to DARPA. Twigg was perfect for the part, always looking as if he were laughing at some private joke. He was never seen out of a sport coat, shirt, and tie, Smith recalled. "Everyone thought he worked for us, for Northrop."

"I had the haircut," Twigg interrupted with the air of a man who always likes to be in charge. "I had the crew cut, and that was the only thing that gave me away."

Most of the wives and many children were at the ceremony. And that illuminated something it was easy to dismiss: that working in secrecy diminished the lives of the workers. It's not just that they couldn't answer the question, "Daddy, what did you do at work today?" but that their family lives could be jeopardized by the black hole of nondisclosure, which could quickly fill with suspicion. Security, one black worker said, was like wearing a lead raincoat.

The question could never be avoided: Was it really the job, or was it something else? A guy could be making it all up because he had a bimbo in Burbank, a floozy in Floral Park—hell, a whole second family someplace, or a bad gambling habit, or an unsavory job with organized crime. There were cases of con artists who pretended to be working in Dreamland, or some other secret facility, or for the Skunk Works.

Was there any sharper symbol of the isolation of the black world than the "hello" phone, the one-way dead-end telephone number given to families of those working at secret facilities like Dreamland?

In *Blue Sky Dream*, his memoir of growing up with a father who worked in SAR programs, David Beers gives a child's viewpoint of all this. His father worked on Star Wars projects for Lockheed's missile division, near the mysterious Blue Cube, the spy satellite control center in Sunnyvale, and during the eighties was dispatched to a place that may very well have been Dreamland.

He was gone for days and weeks to a place the mysterious people on the phone called The Ranch.

"Hal there?" an extremely serious male voice would ask whoever picked up the receiver at my parents' house.

"No, can I tell him who called?"

"Tell him Gunner called. From The Ranch. He'll know."

What was this Ranch where Ronald Reagan had created new work for my father and for "Gunner" and for how many more? My mother and her children were curious, of course, but we had only the slimmest of details with which to construct a mental picture. We knew a man would find himself in some very high and precarious places at The Ranch, because one time my father returned wearing a strange pair of glasses, clunky plastic frames bought off a drugstore rack. He had lost his, he said, "while stepping onto a catwalk. I bumped my head and off came my glasses. I heard them hit the floor about, oh, eight to ten seconds later." My father smiled as he said this, smiling as he tended to smile when he had just told you something that was very intriguing but just shy of violating his security oath.

We knew The Ranch was a place that could be very dark, because another time my father came back with a scabbed cut in his forehead. All he would tell us is that he had been driving across some dim landscape in the middle of the night in a rental car with the lights off and he had run into something and his head had been thrown forward into the steering wheel. "Why were you driving in the dark with no lights?" his wife and children wanted to know. But his answer was a smile.

To spy, you must agree to be spied on. To create a spy plane, you must agree to have your phone tapped, take lie-detector tests, have your background and clearance reviewed every five years.

That was the cost of working in Dreamland. Indeed, this seemed to be the key reason for the ceremony and the revelation of Shamu. At last the wives and the children, now grown, could be told. On black projects such as Shamu, Smith said, "All normal methods of communication are avoided, all identities and relationships are denied. Total isolation is the goal, and this caused hardship." Divorce rates are high in black projects.

When the Air Force first began operating the Stealth fighter at Tonopah, it flew only at night. As a result, the pilots slept during the day. When they returned home on weekends, they would either continue to sleep all day, ignoring their families, or try, usually in vain, to switch their sleeping schedules, leaving them groggy and irritable. Some said they felt like vampires. Many pilots complained of a nagging exhaustion they could not shake. One of them was Ross Mulhare, who died in July 1986 when he flew his Stealth fighter into a hillside near Bakersfield. Mulhare's family did not know what he was doing during the days he disappeared into the desert south of Tonopah, but they did know that he had to take a lie-detector test every three months.

Those who work on black projects must sign an agreement to respect the secrecy of information protected within Special Access Programs, called Sensitive Compartmented Information. These agreements, which for earlier programs were carried out under Reagan-era Executive Order 12356, involve an explanation of the system and "indoctrination." Those inside understand they can be punished—fined and sent to prison for years—under sections 793, 794, 798, and 952 of Title 18 of the U.S. Criminal Code.

Secondhand accounts of the black world abound with tales of persuasive briefings punctuated by shouting and the near proximity of the muzzle of an M-16 rifle to the subject's face. You will disappear, they are told. One former Red Hat flier simply took it for granted that people who talked about the program would disappear.

But the real teeth of the system, the tools for ensuring secrecy, are much more mundane: the threat of the end of a career, of loss of a pension, the regular administration of polygraph tests, the monitoring of phone calls and mail, the careful registration and tracing of the disposition of controlled documents and computer files. The Office of Special

Investigations or FBI may also tap phones and watch the movements of employees and even family members.

With these tactics it is much easier to keep secrets than one would think. At first the black world was a world of intelligence—information. But beginning with the Manhattan Project, black methods were applied to the development of hardware—not just knowing things but building things. The Western Development Division of Air Research and Development in 1951, the first U.S. effort to develop an ICBM, was another early black program. Funding for these comes from budget lines with code names or vague headings. Some, like the U-2, were funded from various CIA funds. But the CIA is only one of some thirty-eight U.S. intelligence agencies, departments, and divisions, and its $2 billion budget is dwarfed by that of the National Security Agency.

Today, entire categories of operations are black as well: the SIOP (single-integrated operating plan) for fighting a nuclear war, "continuity of government" plans following a nuclear war, or antiterrorist operations, for instance.

The biggest misunderstanding about secrecy is that it is a matter of levels, that higher clearance gives one access to more stuff. In fact, the key is not vertical but horizontal—in compartmentalization. The engineers building a Stealth fighter are separated from those building a laser weapon; being cleared for one highly secret project does not mean access to another.

For this reason, the black system was developed with almost scholastic rigidity. Beyond such commonly known stamps as Top Secret or Classified are code warnings like WINTEL: Warning Notice—intelligence sources and methods involved; ORCON, originator controls access and distribution; NORFORM, meaning not to be seen by foreign nationals; NO CONTRACT, meaning not to be seen by contractors.

Categories of information had names different from those of the sources of that information, as part of the compartmentalization process. Such names are almost a parody of themselves. Readers of John le Carré will be familiar with the use of separate code names for a body of intelligence information and its source. The material called "Witchcraft," for example, is produced from a source called "Merlin."

In the fully developed Cold War system, categories of intelligence had names like Umbra and Spoke. Gamma was the name for intercepts of various Soviet communications. (It was also applied in 1969 to the program of spying on American leaders in their protests against the Vietnam War.) A whole host of "G" words—Gant, Gabe, Gyro, Gut, Gult, Goat—some real words, some made up, were used for specific categories of these intercepts: Gamma Guppy, for instance, was the name for overheard telephone conversations of Soviet leaders being driven around Moscow in their limos. It seemed to consist largely of gossip about their various mistresses.

Secret hardware programs received special names, like the Byeman names for spy satellites. The U-2 was Aquatone and Idealist. Discoverer covered Corona, the first spy satellite. But it was in the naming of research programs like Teal Rain and Have Blue by the services, and by such agencies as DARPA, that the new tone of the black world emerged. Something else works to protect secrecy: a sense of fraternity, the qualities of a secret society, a sense of belonging to something special. ("Special" is a key word in the Pentagon. "Special weapons" are nukes; "special operations" are commandos.) To define a group, a cult, a religion, not only are certain key words used but certain words are not used. In the black world, there are terms that are never spoken aloud, like the true name of God. You never say Groom Lake—you say "the Ranch," or "the remote location." And rarely do you even say black.

———

There were also active efforts to penetrate security—like LeMay's old security testers in SAC—and others listening in on family phone calls and watching employees to see that the penetrators were not succeeding.

"There were many efforts to do this," one of the Whalers told me, adding proudly, "To my knowledge, none of them were successful."

"Were there also," I asked, "active disinformation efforts or cover stories?"

"You'd have to ask the professionals about that," he answered.

The professionals. AFOSI? FBI? When I did ask them, of course, I got the inevitable "We can't talk about that."

———

Besides the little whale lapel button, many of the whalers wore another pin: a diamond arrowhead, icon of the Pioneers of Stealth, the loose organization of black-world engineers who had worked on Stealth and now met for occasional reunions in a wave of nostalgia for those early days.

To the Interceptors, all this lent the hope that more craft that the Air Force and contractors had been hiding might soon emerge, other unidentified flying craft, like the Manta or Aurora. Like Q or the Tier III, which might or might not be the same thing. Q was said in the Lore to stand for "quantum leap in technology." It was also a traditional, even legendary designation for the top security clearance. Q, depending on which tales you believed, was either a successor to the failed Aurora or its code name. Or was Aurora a cover story for Q?

I hung around several of the pioneers, and eavesdropped as they spoke of their next reunion. Two of them were talking about the conclave. I gathered that an invitation would not be forthcoming. They discussed who might be attending. Several names were mentioned, then one asked, "And who should we invite from Q?"

———

Before the ceremony, I had walked around the base and the museum, trying to understand how Tacit Blue fit into the aviation history laid out there like a diagram. I was surprised by how open and green the place felt. I had not realized that the base was built around Huffman Prairie, the Wright Brothers' flying field. Today Wright-Pat is huge, three airfields in all, and it is the center of the Air Force Systems command, the MIT and CalTech of aviation high-tech. It is also the home of the Foreign Technology section—perfect for investigating captured MiGs and, as the youfers believed, wreckage from Roswell or other saucer crashes.

Even if there is no Hangar 18, no "level 5" where the Roswell bodies are supposed to be on ice, plenty of buildings here looked right for the part: odd tanks and pipes, cubes and spheres, weird-shaped wind tunnels, all decorated with wisps of mysterious vapor.

20. The Anthill and Other Burlesques

The myth of Hangar 18 in Dayton would continue to grow. No report on it has failed to mention how Senator Barry Goldwater was not allowed access to the building, even with his top clearance. It was well known that he had tried to get Curtis LeMay to let him in. On the Internet, Robert Collins had recently posted an elaborate report on "underground vaults at WPAFB" based on inside sources and infrared photography.

At Wright-Pat they find all this exasperating. They receive inquiries daily. When Frank Kuznik, reporter for *Air & Space* magazine, visited the base for a story, he found irritated scientists tired of dealing with the inquiries. No, there was no Hangar 18 or any vaults full of bodies. But if there were, "Do you think we'd tell you? Don't you think we'd be able to hide it?" Another scientist declared that he wished they did have something alien to put on display, because at just a dollar a head they would certainly make enough money to solve his budget problems.

Long before Area 51 meant anything, Hangar 18 had seeped in to popular consciousness. Now Area 51 was becoming a larger version of Hangar 18. But around Dreamland, deeper, darker vaults were suspected.

John Lear held that the Skunk Works had moved from Burbank to Tejon Canyon, the Northrop radar cross-section (RCS) range west of Palmdale, the better to hide sinister projects. To him and to others, that facility was "the Anthill," where aliens ruled, incubating hybrid humans, gathering abductees for their vital enzymes. Deep in the night, these watchers say, the portals open to emit flying saucers from structures beneath, extending five, ten, even fifty stories below the ground. According to Gary Schultz of Secret Saucer Base Expeditions, the erstwhile self-appointed expert on Area 51 and the leader of regular trips to the perimeter, "We have found out with incontrovertible proof" that the RCS is only a cover, that the Anthill has forty-two levels underground. Things come streaming out of there at night, the tales went. There are reports of mysterious blue beams and "surveillance orbs the size of basketballs."

Schultz had flown over it in September 1991. He had seen evidence of concrete being poured twenty-four hours a day for weeks—a million cubic yards of concrete. The skeptic will ask, Where were all the cement mixers lined up? And where was all the dirt?

Those who believed in the underground bases suspected not only the Anthill but all the RCS facilities. These strange installations look like Dreamland *should* look but doesn't: They have mysterious concrete tilting walls, diamond-shaped pads and panels, shadowed overhangs, James Bond–like facilities of the sort that leap to the imagination at the very utterance of the phrase "Area 51." They are the radar cross-section facilities of the western deserts, the local chapels of stealth, landmarks of the Greater Dreamland: Gray Butte, Tejon Canyon, Helendale, China Lake, White Sands. Hey kids, collect 'em all! And the Interceptors did—they would make trips to each of the facilities. Tom Mahood even tracked down their ownership in real estate registers and public records.

From the air, they are especially sinister, their runways painted with the warning, RESTRICTED RUNWAY DO NOT LAND.

There is a similar facility—perhaps the largest—inside Dreamland, behind the base itself.

RCS facilities test how difficult it will be for new aircraft to be seen by radar. Each major aerospace contractor has one. In such facilities, models of new aircraft or missiles are set on pylons and test radars are beamed at them. Other aircraft might fly overhead, testing their onboard radars. The tilted walls contain and control the deployment of the radar waves. Engineers measure the way aircraft reflect radar beams, how much and in what direction.

To protect against overflights by nosy satellites, some of the models can easily be moved inside walls with sliding doors or have covers quickly placed over them. Such is the most practical and banal explanation of the facilities. To those less trusting, the RCS sites are openings to underground bases, portals to an underworld of secret treaties and alien takeovers.

To the two leading underground theorists, Richard Sauder, author of the aggressively titled tract *Underground Bases and Tunnels—What Is the Government Trying to Hide?*, and William Hamilton, even Plant 42 in Palmdale had secret floors beneath it. In his book *Cosmic Top Secret* and in the video *Underground Bases*, Hamilton claims that the first saucer wreckage came to Area 51 in the late forties and the first underground labs were built at that time. They have expanded ever since. Hamilton describes baseball diamonds and swimming pools that exist beneath the surface.

Hamilton is not alone in this conviction. Even such a fairly sober youfer as Stanton Friedman was intrigued by the idea of levels beneath the runway. "I have been informed that a secret underground base was built under the runway at Groom Lake in the early 1950s, well before the U-2 program," he wrote in an Internet posting. "I expect to dig into this one soon."

Hamilton is disarmingly nonfanatical. He could be lecturing a class on post-Keynesian developments in macroeconomics when he discusses secret tunnels. Hamilton cites sources describing a network of tunnels linking bases, a virtual underground interstate system.

For proof, Hamilton and Sauder offer plans for underground command posts and living quarters and diagrams of tunnel-boring machines from a 1959 RAND report or via the Army Corps of Engineers, organizations that, at the height of the Cold War, were ordered to figure out how to put practically everything underground. The fallout shelter fad was about to begin. Living underground was not considered far-fetched or sinister.

Underground is, of course, rich with metaphor, as the place of the unseen, the realm of death, of organized crime, and of defiant resistance. The idea of the underground base as hive or anthill is common—areas are "honeycombed" with tunnels. In *Them!*, the classic fifties science-fiction film and a parable of the Red menace, a little girl who has been terrified by giant ants (the offspring of radiation) is examined by doctors and a scientist. When the scientist gives her a sniff of formic acid, the poison from ant stings, she begins screaming uncontrollably: "Them! Them! Them!"

———

Thinking of *Them!*, I drove west from Edwards toward the place known as the Anthill, cleverly disguised as Northrop's radar cross-section testing facility, west of Willow Springs. From Trader I learned how to get there, driving past shopping centers and the Willow Springs racetrack, bright with painted ads. The trip was a wonderful excuse to go badassing along primitive roads in a 4-by-4, playing rough road rock-and-roll. Once out of civilization, I followed a dirt road named, perhaps ironically, Broken Arrow. Broken Arrow, of course, is the military code name for an incident involving the loss or theft of a nuclear weapon.

Someone had painted a bright blue warning skull on the rock at the turnoff for Broken Arrow Road, beneath the metal sign with its cincture of welded letters. Broken Arrow was a western Mojave road as hard as the iron of the sign. Here and there were ugly ruts and cracks where the road had dried and split open. At other places, gray clay creeks appeared. At one turn, a false trail, I came to a barbed-wire gate and a jackrabbit flattened on the road, in the pose of a leap, as if captured in midair by a strobe flash.

At last the odometer showed I was close. I parked and scrambled up a hill. Over the horizon I could see all there was to see: a couple of buildings, a radio antenna or two, a water tower. No evidence of underground structures. No air vents, no strange doors. All I saw were signs of new water management facilities—canals and culverts. Nothing suspicious, although it was through the sewer system that the ants in Them! had raced most dramatically. To the suspicious, such innocent stuff was the whole point: The underground was a version of that oldest of menaces, the unseen.

The underground can also be understood as the unconscious—the source of dreams and psychoses. To those who believe in the underground bases, this analogy is more specific. To them, physical levels are indications of levels of information and security, and also perhaps of psychic levels: The deeper the facility is dug, the deeper the conspiracy. If a vision of things below the surface represented the "cover-up" in literal form, connections among them represented the extent of hidden links. Tunnels, the theorists argued, tied the sites together—sinister hidden connections made manifest. The accounts included stories of workers who had ridden the rails from the beach in Los Angeles to Area 51, with connections available to Los Alamos and Sandia. Were Amtrak so well run, it would put the Japanese bullet train and the French TGV to shame.

Thus Area 51 connects with Edwards and Sandia and Los Alamos, and even with the most terrifying of the projected underground facilities, Dulce, on the Archuleta Mesa in New Mexico. Level 4 is concerned with telepathy and dream control. Level 6 or 7 holds the vats with the embryos of half-breed human-aliens and other grotesque genetic experiments. It is known in the Lore as Nightmare Hall.

The underground conspiracy buffs tend to equate security levels with physical levels. Twenty-four or thirty-eight levels of underground installations correspond to the same number of levels of "clearance." But in the actual black world, it's not just a matter of higher or lower clearance from sensitive to secret to top secret to "Q" but of separation on the same level: of different rooms on the same floor. In reality, there is not only distinction among levels but distinction among rooms, so to speak, at the same level.

In their descriptions, the Lorists seem especially concerned with doors. As if they were film production designers, they describe in detail access panels, sliding cards, retinal readers, weight-triggered access doors. Many door controllers or speakers are in the shape of an inverted triangle. The inverted triangle is linked in other parts of the tales to the trilateralists and, more implicitly, to the existence of layers below the surface: It's the inversion of the pyramid on the dollar bill and the great seal of the United States.

The end of underground theories is to see the earth itself as hollow, to imagine not just a hell beneath our feet but the world as a mere shell. In this ultimate version of conspiracist theory, Nazis fly the saucers they have developed into the center of the earth through hidden portals at the poles. "Commander X"—former "Military Intelligence Operative" and author of *Underground Alien Bases*—has the Nazis colonizing the center of the earth in cooperation with "Serpent People" aliens. He believes that "in reality, many of the craft seen over Area 51 in Nevada are not constructed by aliens. They are instead experimental vehicles derived from the secret plans of German scientists, many of whom were brought to the U.S. and given political asylum, even though they may have taken part in vicious war crimes." The Nazis perfected anti-gravity and time-warp transportation, X also tells us, and landed on the moon before 1945.

Hollow-earth theories are as old as the Egyptians, of course, but as recently as the nineteenth century they were taken with some seriousness in the United States. The hollow-earthers populated the center of the earth with all the features and creatures later theorists and science fiction would transfer to other planets. Before there was an expectation of space travel, the interior of the planet was the most distant region imaginable. So *A Journey to the Center of the Earth* would give way to *A Trip to the Moon*.

In 1819, John Cleves Symmes propounded his theory that our hollow earth contains five concentric lands. James McBride explained it all in the following decade in *The Symmes Theory of Concentric Spheres*, "demonstrating that the earth is hollow, habitable and widely open about the poles." One writer of the time imagined the land inside as "a white land," full of the whitest of humans. In the 1830s an odd character named Jeremiah Reynolds began promoting a South Pole expedition to prove Symmes's theory. Amazingly, he prevailed upon the U.S. government to fund not one but two such expeditions. One result was the production of very useful marine charts of the southern waters. Another was to inspire Edgar Allan Poe to write such stories of possibly hollow worlds as "MS. Found in a Bottle," "Descent into the Maelstrom," and *The Narrative of A. Gordon Pym*.

To see the earth as hollow was ultimately a vision of profound despair. It meant we literally could not trust the ground upon which we stood. It meant life itself was empty. Edward Shils wrote of "the torment of secrecy," the pain of those who believed that all history took place behind a veil of some kind of conspiracy, that the real motivational forces in the world are unseen, perhaps undiscoverable. This is a hard philosophy to live with.

———

Helendale, in California, the largest of the RCS sites, is the newest such facility. It is huge, with its own runway, near which Aurora was thought to have been spotted. Its main radar area, called, sinisterly enough, the Upper Chamber, seems to cover acres of concrete.

To reach it, I cut through from the highway that ran east of Edwards AFB, then drove over the white sandy bed of the erstwhile Mojave River, past trailers and little houses. I could see the distant hangar—Lockheed yellow—and turning up the road, I came to the fence and gate that barred the way.

To the left of the gate, shoved up practically against that fence, there was a place called Exotic World, a sort of museum celebrating burlesque culture, the home of an old stripper who has collected the G-strings of the great strippers of the past.

Months later, as I sat watching one of those offbeat local-color features TV news loves so much, I thought I recognized a beat-up little trailer and nearby fence. It was indeed Exotic World. There was a nice sound bite from the owner, herself a former stripper: "Striptease was not invented," she said, it just happened, when someone caught a glimpse of a dancer pushed out onto the stage too soon. "Striptease," she said, "is a phenomenon, and phenomenons are not made, they just happen."

I thought of the Air Force general who declared that a secret aircraft should reveal itself only gradually and seductively. Striptease was about imagination more than revelation, and so were the RCS sites: Phenomena just happened there, too. I wondered if a visit to Exotic World didn't say more about the workings of secret aircraft than standing on the concrete of Helendale's Upper Chamber.

21. Space Aliens from the Pentagon and Other Conspiracies

On his way to the Oklahoma City federal building, the bomber Timothy McVeigh slept in room 25 of a motel in Kansas named Dreamland. I took this information as a token of just how closely the fascination with a New World Order, a new political view of the world, was taking hold of the views of Dreamland.

The New World Order theorists had rapidly developed their own lore, decrying the influences of the United Nations and the Federal Emergency Management Agency (FEMA). Black helicopters and white (UN) personnel carriers were making furtive appearances. They were UFOs of the militias.[1]

On the Internet, the theorists reached such filigreed detail of conspiracy that one story even claimed the NWO would abolish all but a single chain of fast-food restaurants: Taco Bell. Believing this, who would not take up arms against the menace?

NWO lore was overlapping UFO lore. On Long Island, in 1996, Ed Zabo, an aerospace electrician, and John Ford, head of the Long Island UFO Network, were charged with attempting to poison a county Republican chairman by slipping radium into his food. Zabo, a government inspector at the local Northrop-Grumman plant, believed that the county government was conspiring to cover up evidence of UFO landings, which among other things had resulted in extensive forest fires on Long Island the previous summer. The district attorney shook his head and opined that "this all convinces me that there is a side to humanity that defies definition."

George Bush and the speechwriters who popularized the glib but murky phrase "New World Order" to label the era that succeeded the Cold War could hardly have imagined that it would come to denote so readily such a malignant mythology. The phrase became a cipher, a place-holder, a linguistic Groom Lake waiting to be filled with speculations. I took it as a sign of the end of the Cold War, which left a yawning vacuum of uncertainty. We missed the Cold War. And I regarded the NWO's most fervent adherents as victims of a kind of post-traumatic stress disorder. It's not easy to take away an enemy you've lived with for nearly half a century. How much easier to deal with an invented enemy than with none at all; how important to the conspiracist for the world to possess an order, even if that order is dark and hidden.

The Interceptor known as the Minister of Words believed that the appeal of this dark mythology was a sign of economic distress. "The un-educated shitkicker class in this country is dead," he argued. However prosperous America seemed in the nineties, life had gotten tougher for the guy with a trailer and a pickup truck.

But some of the early believers in the UFO cover-up were converting to a still darker view: that an even more sinister conspiracy was behind the use of flying saucers in order to drive us into the arms of the New World Order.

———

In the summer of 1996, I visited the national convention of MUFON, the Mutual UFO Network. It was held at a North Carolina Holiday Inn, with the same tone of seriousness and self-fascination as a regional gathering of insurance salesmen or plumbing supply vendors. I noticed that no one smiled.

In one room, on acres of tables, every stripe of UFO thinking was laid out in books and videos. I felt compelled to browse something called *From Elsewhere: Being E.T. in America*, about the experiences of a man who felt he was an alien on Earth. You could also buy mugs and T-shirts and glow-in-the-dark alien sculptures. But my eye was caught by the cover of a book that pictured a strange pentagon and star device and the title *Space Aliens from the Pentagon*. It bore the subtitle "Flying Saucers are Man-made

Electrical Machines. Revised and expanded Second Edition Creatopia Productions™, by William R. Lyne." Cover lines: "Does the CIA write Movie and TV scripts about 'aliens'? Have you been brainwashed: Does the CIA control Hollywood and TV? Did you know the flying saucer is the best-kept energy secret on earth?" The cover art showed the Pentagon as a maze. Inside it was set a swastika and an all-seeing eye, like that on the seal of the United States or a Jungian eyeball in the sky.

Lyne argues that the saucers were faked by the Pentagon or some se-cret group beyond and behind the Pentagon. He writes that "my 'space aliens' are actually people, whose philosophy and bizarre masquerade are alien to the American way of life, since they believe in government by anti-democratic hoax, to maintain the secret power of the Trilateral Commission elite, to whom our lives are very cheap. I am striking back against an 'alien system' which has attached itself to the nation which our ancestors strove to create, which would be invulnerable to the 'aliens' . . . I have concluded that a Secret Government has watched me, attempted to control me . . ."

Lyne turns the Cosmic Watergate on its head. Far from being a cover-up, he asserts, the saucer stories were all a put-on—a Hollywood pro-duction to frighten us into the arms of the NWO, to create Reagan's unifying alien threat. The saucers came not from other galaxies but from Earth. The Nazis had taken the technology of Tesla and developed flying saucers, which they used to fly to exile in South America—perhaps even Antarctica. Werner von Braun flew the flying saucers out of White Sands after the war. Hitler escaped from his bunker to South America and vis-ited San Antonio, Texas, in 1967 as a guest of LBJ. But when Lyne—and he alone apparently—recognized Hitler and Eva Braun, they were quickly hustled away.

Lyne was always a key player in the dramas he described. He told how he had quarreled with Sargent Shriver over his dismissal from the Peace Corps and how in 1975 George Bush had offered him a high po-sition with the CIA, which he rejected.

Lyne's biography states that he saw his first UFO as a child in Kermit, Texas. He received an MFA in "studio arts" from Sam Houston State University; he certainly had artistic talent. His book is illustrated with obsessive and skillful drawings, part engineering diagram, part R. Crumb.

He believed the National Security Act of 1947, dividing the armed services, was treasonous, and that the Roswell incident was a hoax. The "aliens" were dead monkeys from the rocket tests at White Sands, crudely disguised. He had seen photos of them, but they were stolen by a former girlfriend. He produced drawings from memory, in his skillful but jittery style.

He rolled all the myths together, all the government cover-ups into one all-consuming conspiracy. Hitler escapes, and strange artifacts float around in the hands of old Indians in the Southwest. No topic was too large to bring into his web—Lyne delivers a long excoriation of "Platonist epistemology"—and none was too small—the powers in Detroit conspiring to squash the small, inexpensive Crosley automobile of the late forties.[2]

———

While the account in *Space Aliens from the Pentagon* possesses a singular viewpoint—all the information had somehow come to Lyne and Lyne alone—another perspective on Dreamland employs a dizzying collage of clippings and reports.

In two videotapes entitled *Secrets of Dreamland*, a man named Norio Hayakawa, who had led the Japanese TV crew to Bob Lazar, had produced a carefully, not to say obsessively, documented depiction of a vast conspiracy swirling about Dreamland like a dust devil.

The tapes are made up mostly of footage of a lecture Hayakawa had given to a religious group called the Prophecy Network. He makes token gestures to an apocalyptic sort of Christianity—probably for the benefit of the audience, whose favorite book of the Bible is the infinitely interpretable Revelation. The lecture is generously illustrated with clips about exotic military programs for mind control, electromagnetic warfare, lasers, and exotic aircraft from such sources as *The Washington Post*, *The Wall Street Journal*, and *Aviation Week*. The lecture is followed by home-video footage of flying saucers along Mailbox Road: lights jumping in the sky and turning on the proverbial dime.

UFOs, Hayakawa concludes, are part of a created threat designed to stampede the populace into accepting the New World Order. He reports, "Dreamland is said"—that passive tense again—"to be an acronym for Data Repository Establishment and Management Land. It will be the center for a future satellite linkage system that will centralize all global computer data network systems.

"A device known as Battle Engagement Area Simulator and Tracker (B.E.A.S.T.), developed by the U.S. Naval Research Laboratory and [to] be launched into orbit under the auspices of DARPA, will link all global data network systems in the air.

"The Beast"—yes, the noted Beast of Revelation—"will be some type of a super-computer linking station launched into orbit in a few more years. It may link stations emitting hologramic images into the atmosphere to control the 'thinking' patterns of the populace."

In the hours I had spent watching Hayakawa's *Secrets of Dreamland* videotapes I had noted his shift toward the conspiratorial. There were two tapes, released a couple of years apart, and between them was a subtle change in emphasis, extended even to the packaging, and an apparent shift in his target audience from the youfers to a New World Order conspiracy audience. It was not only good marketing, reflecting a changing world, but indicated a change in Norio's thinking. He believed that the Rockefeller Foundation in North America and the Rothschild financial conglomerate in Europe are an integral part of the entity known as the Bilderbergers, which plans to establish the New World Order by the year 2002.

"The Lord," Hayakawa announces, "is literally coming to catch his believers in the air. A mass confusion will take over the world." I wondered at that moment whether he had ever heard the Louvin Brothers' song "The Great Atomic Power," in which the victims of the A-bomb rise to meet their savior in the air.

His argument draws equally from scripture and *Aviation Week* and goes like this:

The New World Order, a secret government, is using UFOs to frighten us into accepting their tyranny. Strange new technologies are controlling us, including holographic projection and other forms of mind control.

"It is my opinion," Hayakawa insists, "that an elite group of globalists has always believed that the ultimate way to create some type of global unity was to create an artificial threat from elsewhere. It could be war, disasters, worldwide calamity, et cetera, to create an artificial 'crisis.' But the ultimate one is to create an external threat from 'outside,' and the most convincing one will be an 'alien' threat from beyond earth.

"To this end," he intones, "I believe that we have slowly been brainwashed and manipulated to believe in the existence of 'extraterrestrial' entities. Look at the proliferation of 'alien'-related films and TV documentaries and semidocumentaries. I think that this is all a part of the conditioning process that is preparing us psychologically to accept the 'alien' presence and sensitize us to the 'alien threat' in the very near future."

He talks of devices to control minds, some of which may cause temporary memory loss. Certain chemicals are used, and equipment. There is reference to a Dr. Igor Smirnoff—very much his real name—who developed an acoustic device for mind control. Work is going on at Wright Patterson Air Force Base to create brain-actuated airplane controls.

Hayakawa delves into some Joseph Campbell–like interpretations as well. The legendary Majic or MJ-12 from UFO lore is traced to symbolic code words, an occult term from ancient days, linked to magi, or wise men.

The secret government is sensitizing us, he says, preparing us for the takeover. The clips from the popular press prove this. "When The Washington Post says so, it is already done."

Hayakawa narrates the video clips of his saucer-chasing expeditions that follow his lecture, like an appendix in a book, in a very different voice. "The intensity of sound stunned us," he says breathlessly in one. "You could physically feel the noise from eighteen miles away." There are shots from Freedom Ridge, a bouncy, smeary view of the base at night, and a red glow. Is it a plane? "It might just be a car," says a voice on the sound track. "No," another voice, overflowing with excitement, counters. "That's a ship. See, there are trucks around it? . . . They're getting ready to send it up."

———

Because of these videotapes, I ended up one August day in Los Angeles's Little Tokyo. I stepped from the heat into the cool dark lobby of a Japanese American funeral home. It stood near a toy warehouse in an area not so much ethnically colorful as ethnically triumphant: architecture as slick and corporate as Tokyo's, a Buddhist temple in its own little park, a series of looming brutalist apartment buildings with mall. At one edge of Little Tokyo stood a replica of a building from the internment camps, a tattered barrackslike building that might have been pulled from the wreckage of an abandoned training base—the old Tonopah, say, or Indian Springs.

I waited in front of a sign that read SLUMBER ROOM VIEWING. A sweet odor filled the air, and somber Japanese Muzak drifted by. Then a friendly man emerged: Norio Hayakawa, UFO buff, Area 51 researcher, and full-time funeral director.

I had e-mailed Hayakawa, asking to talk to him, and he agreed. He delicately warned me not to mention UFOs if I called him at the funeral home. "You know how it goes," he said, in the tired phrase of many saucer buffs trying to get by in the more mundane world.

I wanted to hear how Hayakawa would tie Dreamland into the Book of Revelation, a dangerously heady elixir for preachers and prophets of many shades.

It turned out to be a little more complicated: Hayakawa went easy on the specific biblical references, hailing a more general "spirituality" that, along with the unification of the various militias, he sees as our best hope of salvation.

After the teeming conspiracy tales of the tape, I hardly knew what to expect of Hayakawa in person. He was gracious, friendly, disarming. We drove to a restaurant on the edge of Little Tokyo. He was honored by my visit and interest, he told me, but he seemed weary, tired of it all.

"My main thesis," he pronounced, almost as if by rote, "is that highly developed technology could be utilized to stage a fake alien invasion to desensitize us to intrusive authority and shocking revelations.

"I think it's always going to be a mystery. It will never be solved. Or by the time we find out what is there, it will be too late. We won't find out until all hell breaks loose."

In his lecture, Hayakawa points out that the year 1947 was when all these strange things began to happen: the founding of the Air Force and the CIA, the Roswell crash. He does not mention the death of Bugsy Siegel and the bankruptcy of the Flamingo, the transistor or the Truman Doctrine or Yeager's first flight through the sound barrier. When I asked, he explained that he traced his own fascination with what he called "the UFO phenomenon" back to that year, perhaps because in 1947 his father, a fisherman, looked up from his boat off the coast of Japan and saw a strange light in the sky.

Hayakawa graduated from high school in Yokohama, and joined Japanese UFO groups in 1963. He attended college in New Mexico and made contact with other UFO watchers in the state. By 1976 he was teaching at a school in suburban Phoenix. In 1988 he watched with fascination when the Fox network, already working its reputation as the tabloid of TV, broadcast the show *UFO Cover-up? Live!* It was the first time Hayakawa heard the term "Area 51."

Nearly a year later, he heard Bob Lazar speak on the Billy Goodman radio show. Hayakawa had long served as a kind of UFO scout or consultant for Nippon TV in Japan, and he let them know about Lazar. In February, NTV sent a correspondent and crew to Las Vegas with Hayakawa.

"Lazar showed us the documents concerning his work," he explained. "Later we found out that his Social Security number"—on the famed W-2 that showed Lazar being paid $977.11 by Naval Intelligence—"belongs to a person in New York."

On Lazar's advice, Hayakawa and the crew headed up to Mailbox Road to look for saucers. "We were looking toward S-4, over the Jumbled Hills, when this strange light came up, went up and down. It was one of the most amazing things I've experienced."

Hayakawa's video shows jumpy images of lights. You can hear the hissing, flickering sound the desert wind makes in the microphones, threatening to obscure the signal.

Hayakawa made many trips back to Mailbox Road, sometimes with Gary Schultz. He had seen the saucers, he thought, and on his tapes there are many lights moving erratically in the sky. He also believed he had seen UAVs flying up there. But he had also come to wonder if some of the things flying had not been illusory images, projected somehow, perhaps holographically—high-tech illusions.

Today, Hayakawa says, "mystification" has taken over Area 51. He sees patterns and connections everywhere; he links the Beast computer to the Book of Revelation. He is fascinated by numbers. Did I realize, he asked, that 1998 was a critical year, 666 times three? Something big would happen. Is it all part of the symbolism for a diabolical trinity? he wonders.

He believes that Lazar's claims are so far beyond any verification that he feels he has been a tool of disinformation—and quite likely mind control. "I believe he was used unwittingly to spread disinformation."

"High strangeness," he said, as if in conclusion, "high strangeness." He sounded as if he felt betrayed. I sensed he was a bit weary of the contention, even weary of his own theories. He was, I felt, the outsider par excellence, as alienated as only a Japanese American running a funeral home could be.

At the end of our conversation, country-and-western music somehow came up. Hayakawa finally came alive. He brightened all over. It was amazing to see him finally smile. He was wild about country-and-western music, he said. He had a portable keyboard system, which he had brought with him and played at the Little A"Le"Inn. I suspected that he wanted above all to be a real American.

I felt a wave of affection for Norio Hayakawa, sympathy for his fragmented roles, for his disappointment in Lazar. He had in a sense been left at the altar by the UFO world, embarrassed by Gary Schultz, stood up at the airport by Bob Lazar.

On the label of his video *Secrets of Dreamland 2*, Hayakawa is referred to as a "phenomecologist [sic] and researcher." A typo (so frequent in UFO material) or an effort at new coinage? Perhaps an attempt to combine phenomenologist and ecologist—"the ecology of phenomenon." The very phrase teemed with possibilities. I thought of a course of university study I had recently heard of called "media ecology." In a sense, Hayakawa's work was a mad gloss on the media. All his clippings about secret mind control programs, implanted chips, holographic projections, and the like were perfectly documentable in journals describing the frontiers of research. But that word "phenomecologist" also suggested "pharmacologist." Was conspiracy thinking a drug prescribed for existential ills, an antidote for alienation, the way Hayakawa's music was?

Later, he sent me his demonstration tape. Its cover showed him in front of a pickup truck in the middle of the desert—a classic C&W shot. In the desert, I thought, everyone is a country star, everyone is an American—and everyone is an alien.

His songs were classics—"Branded Man," "Why Me, Lord?," "I'll Fly Away." I would never have expected to hear anything, however, like Norio Hayakawa singing "Sensuous Woman." I had forgotten to ask him if he had ever heard "The Great Atomic Power," but I somehow felt he lived in the apocalyptic spirit of that song.

Hayakawa's tape became a key part of my personal sound track for the desert. With ZZ Top's version of "Viva Las Vegas," it was instrumental in keeping me awake on the long drives around the perimeter of Dreamland.

———

After I talked to Hayakawa, I headed up the road from Los Angeles to Las Vegas. The link between the two cities, it always struck me, was like that between the cartoon sleeper and the dream bubble above his head. Vegas was L.A. distillate, a step further into fantasy than even Hollywood would go. I was following in the footsteps of men who had bigger ideas than L.A. could grasp, men like Bugsy Siegel, the inventor of Las Vegas.

The way rocking in a boat all day means that when you close your eyes at night on land you will feel waves, immersion in Norio Hayakawa's thinking left me seeing links everywhere. I donned his worldview like a pair of polarizing sunglasses. I listened to the radio news with suspicion. I noticed a weird symmetry in the way the sun was going down and the moon rising—the two circles the same height, the same size. Why?

It struck me, trying to go with his flow, that Hayakawa was a Jungian—he saw just as many archetypal meanings in the world. His was a world teeming with meanings—too full of meanings, perhaps. Conspiratorial, "mystified" meanings. "Mystified" was what he had said about Dreamland—"they've mystified it"—and now they were using our fascination with it to delude us.

A little later in the day, when a song by Willie Nelson came on, a verse stuck in my mind, a warning that sometimes your dreams can begin dreaming you.

———

Norio Hayakawa, an admitted "conspiratorologist," seemed resigned to living in the world such a role created. This meant that he encountered Dreamland everywhere he looked. In return for understanding the secret order, he had to accept the impossibility of escaping that order—it grew to take dominion everywhere. "The place has no edges." Hadn't the Minister warned us?

Pulling in for gas at a truck stop in Barstow, I came across the Area 51 video game. It was a big hit in the arcades, I learned, and I kept running across it in diner lobbies, in mall arcades, outside movie theaters.[3]

The game's opening screen summed up perfectly the new pop mystique the phrase "Area 51" had taken on: "Area" in military crate stencils, "51" in big bank-vault metallic letters and numbers, dented with bullets. This was the vague popular understanding of the mythical, imagined "Area 51"—the aliens have literally possessed military bodies. "The fate of humanity hangs in the balance," the instructions explain, at the same time promising a "detailed re-creation of the most secretive airbase in the world." If you read the fine print, you learn that "Area 51" had been trademarked.

The premise of the game is that a saucer, or other craft, has been recovered and its occupants have taken over the base. The player must get inside as part of a special SWAT team and battle "alien-infected personnel," a handy means of conflating evil, nonhuman expendable aliens with traditional images of bad guys wearing berets and overalls—the type that die by the dozens in Hollywood action films. Boxes and barrels surround the place; there are hot fighters and tough humvees scattered about. A panel truck bears the designing firm's name, Mesa Logic.

The premise will come as no surprise: Shoot 'em fast as you can as they pop out from behind boxes and vehicles or dash along catwalks. Hangars make fine settings for shoot-outs. The ultimate goal of the game is to "penetrate" far enough to set off a special nuclear destruction device and rid the planet of the invading scourge. I couldn't help noticing, a little wistfully, that winning the Area 51 game meant destroying Area 51. But when I played, I never managed to get very far inside the perimeter before running out of ammo and lives.

——

With Tom Mahood's detailed time line of Lazar's life in hand, I drove around Las Vegas on my own personal Lazar tour. I passed the WELCOME TO FABULOUS LAS VEGAS sign that marked the beginning of the Strip. I wanted to get a sense of the place where Lazar was said to have "pandered"—the Newport Cove Apartments, site of an alleged brothel. A few blocks off the Strip, I found them: a complex with thick pseudo-adobe walls and the wavy red tile that was supposed to signal Spanish style but looked instead like giant clay lasagna. This was not a dump or a cheap hotel but a fairly high-class if anonymous set of apartments. WELCOME HOME! a sign shouted.

I passed the familiar Glass Pool Motel, one of my favorite places in Vegas, even though it represented a minor gimmick for the Strip, which was in its fetal stages here on the edge of town as if foreshadowing for drivers entering the city the fountains and swim-up bars that lay ahead. But I took it as an early, touching bit of entrepreneurial show business: The pool was raised above ground and fitted with portholes so you could glance in at the swimmers. It reminded me of old-style aquariums, where you could see porpoises through portholes, then go upstairs to watch them leap. I liked it because it still had an amateurish quality to its showbiz, although now the water looked none too blue, murky and uninviting.

I stopped by Lazar's old house, where his first wife had committed suicide. It was empty now. It stood on a nice, quiet street, exactly the kind TV reporters flock to when someone is hauled away on a stretcher, the neighbors telling them that they would never have dreamed of it in a thousand years. The concept "safe house" leaped to mind.

I headed back east past the Janet terminal and came to the edge of the Hughes Industrial Park on the other side of McCarran Airport. This was Dreamland's navel, as it were, the umbilicus connecting it to the real world. The contractors who served Dreamland were clustered here on the map that would double as an organizational chart. There, in neat, slick glass boxes of low buildings, like stereo components arranged in a store, were Wackenhut and SAIC, in the same building as Bechtel. Lockheed sat on its own little loop—Kelly Johnson Drive!—across from EG&G Special Projects.

A gardener was working around the sign proclaiming EG&G SPECIAL PROJECTS. That word *special* again, as in special forces, special weapons, special operations. Having run most of the Nevada Test Site's operations, directly or through its REECO subsidiary, having hired guards and owned aircraft, and now operating most of Dreamland, EG&G had come a long way from the labs at MIT where Harold Edgerton had started out.

———

The man who had invented stroboscopic photography was great PR and beloved at MIT. Harold "Doc" Edgerton's images of bullets passing through apples, and footballs indented by the toe of a kicker, turned technology into showbiz. They reached a wide public, the *Life* magazine sort of audience, and showed science not as equations or test tubes but as something fun and exciting and amazing. MIT president James Killian, who had headed the commission that recommended building the U-2, would coauthor a book with Edgerton on his photographs.

Edgerton's photos also represented a turning point in the way twentieth-century man saw the world. In his standard *History of Photography*, Beaumont Newhall writes that strobe photography had "gone beyond seeing . . . and brings us a world of form normally invisible," which fixes "forever form never detected by the unaided eye." It revealed what art critic Rosalind Krauss would later call "the visual unconscious."

Edgerton's photographs captured the dreams of everyday vision, the moments that slept beneath the waking level of ordinary sight: frozen bubbles and bullets, and the magical crown created by the splash of a drop in a pail of milk.

Born in 1903, Edgerton spent most of his childhood in Aurora, Nebraska, a science-fair whiz kid. In the late twenties he experimented with argon lamps and developed the stroboscopic method of photography, a bright, extremely short flash of light in sync with the camera shutter.

He was fascinated with aviation, having seen the Wrights fly at Fort Myers, Virginia, in 1909; during World War II, reconnaissance aircraft were equipped with his strobes. Edgerton's flash illuminated crossroads and town squares in Normandy the night before D-day, documenting the placement of German troops.

By the thirties it was clear there was too much money to be made with the strobe not to commercialize it. With his key associates, Herbert Grier and Kenneth Germeshausen, Edgerton established a company to commercialize the equipment for industrial clients. Strobe photographs could reveal the inner workings of machines, and, adapted, strobes would pace the party of the sixties—their dreamy lighting inducing reveries while dancing to rock and roll—and sometimes trigger epileptic fits. The strobes later went underwater with Jacques Cousteau and discovered the wrecks of the *Titanic* and the *Monitor*. But their most important use would be in capturing the milliseconds of an atomic explosion, tracking the fireball out from its plutonium kernel, so that *Life* magazine could reveal the unfolding of the nuclear blooms that obsessed its readers.

Edgerton's cameras were at Eniwetok Atoll in 1946 and, a few years later, at the new Nevada Proving Ground, set up on a seventy-five-foot tower seven miles from ground zero. There, they captured the nuclear explosion in the moment it hung like a leukocyte, a terrifying organism blown from micro to macro size.

It soon became clear that triggering a camera to take a picture of an atomic blast was very much like triggering the blast itself, and EG&G became one of the AEC's chief contractors. EG&G didn't just photograph the bombs, it helped to explode them. It produced thyratrons, krytrons, and other detonators. And soon EG&G was running all sorts of things at the test site, such as the building and operation of the blast doors in underground tunnels, which would close in a millisecond.

For the DOE, EG&G developed special bacteria to remove radioactive components from the soil, and it grew top-grade mercuric iodide crystals on space shuttle flights in 1985 and 1992 to serve as the heart of new types of extremely sensitive radioactivity detectors. As the Cold War wound down, EG&G began to look to civilian work. In 1993 it obtained a new contract to manage the space shuttle launch and landing complexes for NASA, a task that, according to the company's annual report, required a "200-man uniformed security force and SWAT team." It also ran facilities for separating tritium—the heavy isotope of hydrogen used in nuclear weapons—from helium. It had branches in Langley, Virginia, in Florida, and in West Virginia.

The company's 1995 annual report listed some $1.4 billion in sales and touted the company's work in sensors for air-bag deployment and other automotive uses, its "Z-scan" airport security system, and other work. There was a terse mention of "continuing assignments for U.S. Customs" and a contract "from a U.S. federal agency to conduct a classi-fied project." However, there was no mention of the Janet airline, or of Groom Lake, or of the decision to let the contract to run the Nevada Test Site go to Bechtel. And there was no picture of the building that houses "EG&G Special Projects."

22. Searchlight

We were heading for the center of the world. In a rented Hyundai Sonata, Trader and his friend and I were driving east from Las Vegas, then south toward Searchlight, Nevada. Trader had read that the Mojave Indians believed a certain mountain called Avikwame was the center of the universe. Based on description and hand-drawn maps, Trader said that Avikwame appeared to be Spirit Mountain, part of the Newberry Range near Searchlight. Trader brought a friend, a journalist who had once accompanied one of Gary Schultz's secret saucer base expeditions to the perimeter of Dreamland.

The town's name suggested a government special-access or black program. I thought of Black Light, the name of one such mysterious program, and Redlight, the alleged secret saucer program inside Dreamland. And the UFO group CSETI (Center for the Search for Extraterrestrial Intelligence) called its program for tracking sightings Spotlight.

Trader tracked the black budget that financed secret aircraft projects. By interpreting the budget's secret codes and mysterious symbols, he had audited the books of Dreamland. He'd followed the money.

Trader was red-haired, tall, not what I'd imagined—he had been for me a stealthy character behind an e-mail name for so long. His real name was Paul McGinnis, and he spent his working hours creating and debugging software for a firm in Irvine—that glittery futuristic planned city packed with aerospace and high-tech firms. He was a "Code Warrior."

It was not by accident that he was also fascinated with mythological symbolism, odd rituals, and the bizarre corners of culture. His extensive home page furnished links to pages on Finnish epics, Betty Page pinups, tattoo art, and voudoun, or voodoo. The site was illustrated with a strange crosslike shape, and Trader explained to me that it was a "vever," a voudoun symbol of the crossroads that was believed to open the gates to other dimensions.

In the budget, Trader looked for the confluences and crossings of information that opened up an understanding of what was actually going on, a search for little vevers in the bureaucracy. In a sense, all Dreamland was a kind of vever, an opening to the black world, linking reality and imagination. You could see the black budget as a kind of hoodoo book of conjure spells, a set of computer viruses in bureaucratic codes—a pattern somewhere between hex and hexadecimal.

He could read along in *Aviation Week*, say, about the specifications of a new airplane, about performance envelopes and flyaway costs, and then all of a sudden the bottom would drop out with a sentence like "The other projects, however, remain firmly concealed in the black world." It was as if you had sailed to the edge of the pre-Columbian map and gotten the message "Here there be dragons."

The black world—and Dreamland itself—was like what computer experts call a black box. A black box refers to circuits or program codes whose functions are known but whose internal structure is not. The internal mechanics do not matter to a designer who uses a black box to obtain that function. Dealing with a black box was a form of reverse engineering, and decoding.

For Trader, it was all about breaking the code, trying to comprehend the inputs and outputs of the black box. It had gradually dawned on me, too, that many people who bought into conspiracy theories, especially those that neatly tied everything together, were engineers or computer programmers, people who worked in worlds where things connected, affected each other, had problems that could be solved. They wanted the rest of the world to work that way—indeed, *saw* the world as behaving that way. They wanted to find the code and debug it.

Trader did for a hobby what intelligence analysts did for a living. He made himself into a collector, interpreter, collator, and on-line publicizer of the black budget and its associated "special-access programs," with code names like Senior Trend and Tractor Bat and Have Donut.[1]

The black budget is the government's classified accounting of the amounts it spends on activities it doesn't want to make public: secret military research and weapons programs, intelligence gathering, and covert operations. It admits of no easy calculation, but Trader guessed it might be as high as $40 billion a year—a figure larger than federal spending on education or health care. Looked at in simpler terms, the government was spending $100 million a day on black work.

He explained that the black budget is documented in funding requests and authorizations voted on by select congressional committees, and published with omitted amounts and blacked-out passages. It hides all sorts of strange projects, not just from enemies foreign and domestic but from the public and their elected officials as well. The Pentagon's black budget is actually composed of two budgets, a Procurement budget and a Research, Development, Test, and Evaluation budget, the tab for the toy testers. There are other black budgets, too, covering defense intelligence and research. The reorganization of intelligence gathering has given us exotic and almost unknown organs such as the Central Imagery Office (CIO) and the Defense Airborne Reconnaissance Office (DARO). An internal Pentagon memo from August 1994 that was accidentally released and showed up in *Jane's Defense Weekly* revealed numbers for some of them: The National Security Agency spends $3.5 billion a year, the Defense Intelligence Agency, $621 million, and the Central Imagery Office, $122 million for spy satellite work.

Trader collects such government documents as the House and Senate versions of the "National Defense Authorization Acts," scrutinizing both the reports and the supporting testimony to Congress. He spends hours consulting the Pentagon's own guides to reading the budget— Department of Defense Handbook DoD 7045.7-H—and with publications like "FYDP Program Structure," Department of the Air Force document "Supporting Data for Fiscal Year 1994—Budget Estimate Submission—Descriptive Summaries—Research, Development, Test, and Evaluation."

These are not exactly light reading, and the plots are slow. Trader soon learned that the black budget was a tissue of truths, half-truths, and quite likely outright untruths, a fabric of disinformation as much as information. Huge items can be hidden by breaking them up into smaller items, mislabeling, or simply omitting them.

Even the names and responsibilities of the agencies involved are often hidden. The National Reconnaissance Office, in charge of spy satellites, was so secret that until a few years ago its very name could not legally be spoken. The "Virginia Procurement Office" is really the CIA, and the "Maryland Procurement Office" is the National Security Agency. And beyond programs marked merely secret are budget items tagged with the wonderful euphemisms "selected activities" and "special access."

Through the Freedom of Information Act, Trader managed to get such juicy documents as the RAND corporation's "Route Planning Issues for Low Observable Aircraft and Cruise Missiles," a manual about the rules for the China Lake airspace. There was also one, he was sure, for the Dreamland airspace itself, R4808N. He had security manuals from the Nevada Test Site that revealed you had to have an "8" on your badge to get into Area 51.

Trader had strong political convictions, to be sure—he supplies politicians advocating reform with inside information. But more than anything, I got the sense he was taken with the joy of the hunt, the thrill of the puzzle.

———

The black budget is the tip of a huge iceberg of secret government records that date back to World War I. Well, not really an iceberg, perhaps, but a glacier of classification, increasingly exposed as the Cold War thawed out the files. The list of odd numbers and funny words that is the budget stands for something more: the true information that belongs to the American taxpayer.

The black budget had its origins in top-secret World War II research like the Manhattan Project. It took on added strength in 1958 in the wake of Sputnik, the Defense Advanced Research Projects Agency, and the use of CIA "reserve funds" for the U-2, the Blackbirds, and other programs. It was the slush fund for Ike's famed military-industrial complex.

Even after the standoff with the former Soviet Union ended, the black budget remained huge. One reason is the Gulf War, which lent high-tech weapons enhanced prestige and strengthened a vision of video-game war in which few human beings—at least on our side—are actually killed or wounded and where information gathering is vital. We fell even more deeply in love with high-tech "silver bullet" weapons.

In a strange way, the cuts in the overall defense budget led to a new emphasis on the sort of weapons for which the black budget is best known. Smart bombs are cheaper than stealth bombers, the argument goes. The black budget may even have increased as a percentage of the overall national budget. By the mid-nineties we were still spending perhaps $20 billion on secret weapons research programs. Some of those programs involved the planes flying out of Dreamland, some were satellites, some were exotic energy weapons. Work continues on mounting anti-missile lasers in Boeing 747s. "You know," Trader said, "Star Wars never really went away."

At work a proud "Code Warrior," Trader would spend long nights trying to decipher code, going through the mind-numbing documents in which the black budget is laid out. He had discovered the black budget because he was a black-airplane buff. Specifically, he became fascinated by Aurora. What distinguished Trader from other Aurora watchers is that he began filing Freedom of Information Act requests about programs whose names suggested they might be aircraft. (Black-budget watchers know that "Senior" is the designation for the Air Force's advanced R&D projects—Stealth was Senior Trend, for instance.) In September 1993, he filed Freedom of Information Act requests for information on what he thought was Aurora—Senior Citizen (Program Element 0401316F)—and on Groom Lake.

Trader found himself exchanging letters with an Air Force colonel named Richard Weaver, then the secretary of the Air Force's deputy for security and investigative programs, and later the author of the report tying the Roswell incident to the Project Mogul balloon.

What really set Trader off was doing an FOIA on the FOIAs he had previously filed: He wanted to understand the process and why his requests had brought back very little real information. Reading his own censored case files, he grew angry. "I became convinced," he told me dryly, "that the Air Force, and other military services, had large numbers of senior officials who held arrogant attitudes towards the average American taxpayer."

In the files were memos from Colonel Weaver recommending rejection of Trader's requests, including such lines as "His appeal 'justification' is the standard [blacked-out censored area] provided by almost everyone else who makes similar requests for this information. All have been turned down. His rationale that he somehow should be allowed to perform those oversight functions of Congress, while novel, is not compelling."

This response turned a mild-mannered inquirer into a muckraker. "I was merely pointing out the Air Force's violations of U.S. classification policy, contained in Executive Order 12356, and how secret spending violated Article I, Section 9, Clause 7 of the U.S. Constitution." He referred to the requirement that Congress approve all federal spending. The black budget, Trader and others argue, violates that provision by hiding the purpose of the expenditures.

He took further inspiration from a book called *Blank Check*, by reporter Tim Weiner, who had won a Pulitzer Prize for his exposé of black-budget programs for *The Philadelphia Inquirer*. Weiner called the black budget "a culture of deception." It is, he wrote, a closed world built on the familiar cozy relationship between Pentagon officials, the military brass, and defense contractors. The result was waste. Weiner had investigated cost overruns and performance failures of programs such as Milstar, the military communications and control satellite. He wrote that it was all about preserving empires, that keeping programs secret is an expression of institutional power, part of the still-closed world of the military and its contractors.

But Trader wanted to go further: into the projects whose very existence was hidden. He began assembling his own black budget, using congressional and DOD documents. It was like reconstructing a crashed airplane or assembling a dinosaur skeleton, with conjectural plaster pieces filling in the missing gaps. He set up an Internet site to distribute his files.

Trader, like most critics of the black budget, argued that for all the triumphs of the Skunk Works, most secret programs hid waste. Revealing the cost of a Stealth fighter tells no more about how to build one than the cost of a Cadillac does. Many black programs, such as the B-2 Stealth bomber and the Milstar satellite system, ended up costing far more than planned, but by the time the public learned of the cost overruns it was too late to kill the programs. So much money had been spent that proponents successfully argued that ending the programs would be a bigger waste.

The B-2 was too big to hide. If the Skunk Works provided stories of how black programs could provide stunning success, the B-2 was the prime public example of the disasters secrecy could produce. I caught my first glimpse of a B-2 bomber one day as I drove past the chain-link at Air Force Plant 42 in Palmdale. It was twilight, and far across the open ground I saw a gray blobbish whale shape, derived in part from what Northrop had learned by flying Shamu. It was the primary example of a black program gone awry. With an undefined mission and an unproven need, pushed by the great momentum of airpower advocacy, its cost ballooned to over a billion dollars a copy, Tim Weiner had calculated, three times the worth of its weight in gold. "It just got away from them," Ben Rich told me, referring sympathetically to his traditional rival Northrop.

The B-2 compounded the cults of airpower and of stealth with a third, the flying wing. In January 1981, a frail, ill eighty-five-year-old man walked into a room at the Northrop offices on Century Boulevard in Los Angeles. Senior officials and engineers welcomed Jack Northrop. From a box they pulled a model of the Stealth bomber. To Jack Northrop, it was instantly recognizable as the heir to the flying wing bombers he had designed in the 1940s. "Now I know why God has kept me alive these last twenty-five years," he said tearfully. Standing beside him were Steve Smith and John Cashen, who had helped create Shamu.

The flying wing had always enjoyed a mystical, almost fanatic following from those who saw it as the pure aircraft shape. It obsessed Jack Northrop. He had worked for Lockheed, designing the Vega and Orion; in his spare time he designed and built his first flying wing and tested it at Muroc Dry Lake in 1929. By 1940, he had his own aircraft company, and his flying wing bomber was approved for construction; its first flight was in 1946. When the YB-49 flew cross-country in 1949, President Truman went aboard. It was featured in the 1953 film The War of the Worlds, looking as strange as the ships that bring the invading aliens to Earth.

But the flying wing was a doomed dream. Northrop tried to modify the prop version with jets, but it lost out to the B-36. Northrop lost his company in 1952, sacrificed on the altar of the flying wing. As the B-2, the flying wing seemed doomed again. Costs rose, and by the end of the Cold War the vision of the B-2 as the successor to the B-52, the B-1, and other SAC bombers seemed absurd. Too expensive and too precious to fly, it sat out the Gulf War.

The Bush administration killed the Navy's A-12 Stealth carrier aircraft before it was ever unveiled to the public. Two billion dollars had been spent—the budget, one journalist noted, for the whole National Park Service. I thought of that every time I saw a photo of the A-12, thought of the lodge at Yellowstone and rangers in little Smokey the Bear hats.

———

Trader's work impressed some of the public-interest muckrakers in Washington who had been looking at the black budget for years. One of Trader's admirers was Steve Aftergood, John Pike's colleague at the

Federation of American Scientists. Aftergood wrote the FAS's *Secrecy and Government Bulletin*, which tracked the progress of those battling excessive secrecy and, in the process, charted the follies of the classification system. It was only a slight exaggeration to say that what Ralph Nader was to Detroit, Aftergood had been to the Pentagon and the intelligence agencies.

Keeping too many secrets is not only undemocratic, he wrote, it is expensive. It requires guards, vaults, background checks. Think of it as servicing the national information debt. A GAO study placed the figure at $2.2 billion, but pointedly noted that its calculations had been hampered by the CIA's refusal to cooperate. Private industry spends an estimated $13 billion more adhering to government security standards.

"The more secrecy you have," Aftergood states, "the thinner your security resources are spread, and there is a loss of respect for the system. That promotes leaks."

Out of incompetence, exhaustion, or spite, leaks had been increasing. The leaks were a sign of institutional decadence, Aftergood explained: "The government has found it easier to let the classification system disintegrate than to establish new standards that command respect and loyalty. If current trends are taken to the limit everything may eventually be classified—but nothing will be secret."

Aftergood described a secrecy structure that might well collapse of its own weight. I got the picture of a crumbling empire with a capital city too poor to keep its walls repaired. The strange, distant civilization of the Pentagon appeared a decaying fortress—Rome with the Huns outside, and the black marketeers inside, trading through gaps in the crumbling walls. In fact, it sounded a lot like the Soviet Union in its final years.

———

We parked under blue skies and continued toward the center of the universe on foot. We climbed into a lovely canyon, its soaring rock walls neatly decorated with green. A few other visitors had clambered up one of the walls and, in triumph, taken off their shirts at the top. The canyon narrowed and twisted and the plants at its base grew larger and more verdant. There was more water deep in the canyon, as well as little beds of dirt where the grass grew almost like a marsh, in contrast to the wide delta of desert into which the canyon opened.

We stopped at a cave where painted bighorns loped across the walls among spirals and concentric circles. The walls were as liberally covered with drawings as a New York subway station. Zigs and zags, circles and slashes, and romping mountains goats and deer. These were homes, and I felt almost like an intruder. They were comfortable little ledges where earlier peoples had slept and eaten and laughed, their ceilings blackened by campfires.

Trader took pictures with his digital camera. He would post them on his Web page, where there was a link to a compilation of Native American petroglyphs in Nevada. I liked the idea that these ancient drawings would be burbling across this most advanced medium as soon as he got home.

He had recently discovered a program for rapidly building makeshift runways and hangars—a program that could turn all kinds of distant spaces into little Dreamlands on short notice. He was looking at something called Timberwind, a project for building nuclear rockets—an idea most people thought had been scotched long ago. Of the Star Wars programs—"directed energy" weapons—there was even less to be found.

———

The next night we went together to a Department of Energy public hearing in Las Vegas. A formal solicitation of democratic sentiment on what the DOE should do with the NTS now that the Cold War was over, it had brought Trader to town. He had studied the eight lilac-covered volumes of the DOE's environmental impact statement, which considered the effects of different courses of action. What would happen if the place was closed? What would happen if it was used for other kinds of testing? But Trader wanted to know why Area 51, which had some of worst known environmental problems of the whole test site, was not discussed. The only mention of Area 51 in the document was this: "Under Public Land Order 1662 (June 20, 1958), approximately 38,400 acres were reserved for the use of the Atomic Energy Commission in connection with the NTS. Management of this land has since been delegated to the U.S. Air Force." This was the old game of shifting responsibility for the place between the Air Force and the DOE.

A hearing such as this is a winning process in many ways, a bizarre and rare membrane in which the public in all its diversity touched the bureaucracy. It made me proud to be an American in a way a flyover by Thunderbirds, for all their powerful engines, high speeds, and amazing precision of flying, did not necessarily do.

These hearings brought out local color. At an earlier one a man had stood up and said, "In the name of God, my name is Moe. I'm a permanent resident who has been living in Las Vegas for over six years. Believe in your God!" With that, he raised his green Koran in his hand and began to speak. The number of the area where the secret base was located was 51, he said, so he would read chapter 51 of the Koran: "Believe in your God. Promise in the winds which blow in holy directions. Promise in the clouds that carry heavy rains. Promise to the angels who perform the orders of God. Promise to all corners that whatever you say is true."

He came to another passage: "Abraham said, 'What is your duty here?' to the aliens, 'What is your duty here?' The answer, 'We are here to destroy the bad crime!' " The man pointed to officials from the Bureau of Land Management and continued: "All aliens! All aliens! We want to see the freedom of those captured aliens, because we are here to save the good from the bad one more time."

This evening was much calmer. There were the usual Greenpeace spokespeople and, in counterpoint, a former Air Force officer who said he had eaten plutonium day and night, and bragged that he "pissed plutonium." He had cleaned up after SAC when the B-52 bumped the tanker over Palomares, Spain, back in the sixties, accidentally scattering plutonium the way the AEC scattered it intentionally in Area 51 as part of Project 57, back in the fifties.

Trader got his chance. He read his formal statement and showed a couple of the gemlike documentary artifacts he had picked up: one, a press release from October 17, 1955, relating to the construction of the Watertown Strip by REECO; the other a letter written on AEC stationery stating that a small private plane had landed on the strip in 1957.

He footnoted a number of references to the place and asked why they hadn't said anything about Area 51. How could anyone make a judgment about the real environmental impact of the Nevada nuclear test site, he argued, if they didn't know about Area 51 and such programs as Project 57 or Project Timberwind (the secret nuclear-rocket program with a classified Environmental Impact Statement)? Both the audience and the DOE panel listened silently. No one seemed shocked by any of this.

The speaker who made the most impact on the audience was a representative of the Western Shoshone, who pointed out that the tribe rejected the whole treaty of Ruby Valley of 1863, under which the U.S. government claimed ownership of the test site. The tribe had never accepted the federal payments that would have put the treaty in force; they still claimed the land. Speaker after speaker had made reference to the fact that taxpayers and citizens owned the test site, and I had always thought of myself as being as much an owner as a watcher. Now I had to consider that Native Americans might own the test site, and Dreamland itself.

Ownership had been important to us watchers. It lent a certain self-righteousness to our demands: We're taxpayers, this is our place, we must be allowed in on its decision-making. But part of the government's secrecy about Dreamland took the form of hiding or denying ownership. At the hearing, the Department of Energy effectively denied ownership, as the Air Force frequently did. But Native Americans, who did not share the white man's sense of individual or collective ownership of land, were now, somehow ironically, claiming it.

From the DOE to Trader to the Shoshone, everyone seemed to be selling a different version of Dreamland, repping their views, agenting their image. They were like real estate agents. I had once been warned of that profession: "The difference between realty and reality is in the I." And the eye.

23. "Job Knowledge"

Driving away from the Las Vegas hearing, I realized that there were very likely petroglyphs similar to those we had seen in the canyon inside Dreamland and in other distant reaches of the Nevada Test Site. There was even a report on the subject—our tax dollars at work. "An Archaeological Reconnaissance of the Groom Range" had been conducted in the summer of 1986, as part of the legal requirements of the 1984 seizure of Bald Mountain and other perimeter areas.

The archaeologists had also found a number of middens—trash heaps—and from the bones and other bits could determine what the tribal people had eaten and how they had lived. In 1994, an effort was begun to poke through Dreamland's own midden. An ambitious and idealistic lawyer named Jonathan Turley, who ran an organization called the Environmental Crimes Project at George Washington University, filed suit on behalf of former Area 51 workers against the Air Force and the Environmental Protection Agency.

The defendants in the suit were Defense Secretary William Perry, National Security Adviser Anthony Lake, Air Force Secretary Sheila Widnall, and EPA administrator Carol Browner. In violation of law, the suit charged, the military at Area 51 had burned hazardous wastes without a permit, exposing workers to dangerous chemicals that made them ill and, in two cases, led to their deaths. The best known, Robert Frost, was a sheet metal worker who had died in 1989 of cirrhosis of the liver before the case had been filed. Frost had tried to sue Lockheed, to no avail. His widow joined the plaintiffs in the suit against the government. Frost had burned waste in open pits, he reported. His skin turned red and began to peel. After his death, a Rutgers University biochemist, Peter Kahn, found concentrations of dioxins and trichloroethylene in Frost's body tissues. Another worker, Walter Kasza, died at age seventy-three of liver and kidney cancer.

Others in the suit gave accounts of burnings in open pits and the huge plumes of smoke from dangerous corrosive chemicals—solvents and sealants, plastics, paint wastes, by-products of composites, and stealth coatings. Their chemical names were frightening even to the layman: dioxins, methyl ethyl ketone, trichloroethylene, and dibenzofurans. The workers were given no protective clothing or masks, they said, even after they asked for them. They were forced to go into the pits and rifle through the half-burned material to be sure nothing was left. Everything was burned—chemicals, papers, leftover prime rib and lobster from the dining hall, furniture, and vehicles. They came down with all kinds of symptoms, not just the skin rashes but eye irritations, headaches, blackouts.

Two big Kenworth eighteen-wheelers were always in evidence, one worker reported, and huge fifty-five-gallon drums were brought in with materials from Burbank. The burning took place at the edge of Papoose Lake, near the storied S-4 of Lazar's tales. Was the Lazar story the military's own bizarre cover for the burning?

The story suggested a pattern like that Joe Bacco had described at the test site, where the sense of national urgency and emergency led to abuse of workers. In 1986, workers for the Skunk Works in Burbank sued Lockheed over illnesses they said were acquired from exposure to substances used in building the Stealth fighter—the chemicals used in its composites and in its radar-absorbing coverings were extremely toxic. The local citizenry had joined in later. That was why the original Skunk Works, the fenced-off wasteland I had visited, was now bare ground. The workers who had dealt with similar substances at Area 51 itself were stepping forward at great personal risk. Even as shielded by the John Doe conventions in the legal documents, they were violating their oaths and jeopardizing their pensions.

As I read about Turley's suit and talked to him, I began to associate the sort of cumulative secrecy Trader had described to me with a great midden packed with layers of detritus. The suits were "citizens' law-suits," not torts or damage claims. The workers weren't after money, they were after information. They simply wanted to know the specific chemicals to which they had been exposed so they could seek treatment. But the Air Force argued that even to take soil and air samples might reveal the materials used in secret projects and thus compromise them.

Secrecy, so useful in crises, could also become a dangerous substance. Turley was charging that the abuse of secrecy was the means of hiding the abuse of the chemicals. The Air Force, he argued, had committed a crime by burning chemicals without a permit, and the result had been the injuries and deaths of the workers. "We have compelling evidence that the government and its contractors have used the secrecy of Groom Lake not to protect national security but to shield the illegal disposal of hazardous waste."

The Air Force defense was that national security considerations protected it even against suits based on criminal activity.

It was a startling and unprecedented claim, far beyond anything Nixon had made at the time of Watergate, for instance. The implications were huge: Would the same national security defense have placed the officials beyond the reach of prosecution for murder? (Two of the plaintiffs had died, after all.) But no one had ever sued a black facility before.

At first the Air Force lawyers denied the existence of the facility, but Turley came up with three hundred pages of references to Area 51 and Project 51 in Air Force and DOE documents, and finally the officials acknowledged the memorandum of agreement that charged them with running it.[1] Claiming that Area 51 did not exist, the Air Force had apparently begun to avoid all references to it, using "Groom Lake" instead.

Area 51, after all, was an obsolete designation bestowed more than thirty years ago by the NTS and the AEC. The Air Force claimed the place was run by the Department of Energy (formerly the AEC), which in turn claimed it had given up authority years ago. The overlapping colors on the map of Dreamland became a means for passing the buck.

By 1994 the Air Force issued a grudging statement of acknowledgment that carefully avoided using the term "Area 51": "There are a variety of facilities throughout the Nellis Range Complex. We do have facilities within the complex near the dry lake bed of Groom Lake. The facilities of the Nellis Range Complex are used for testing and training technologies, operations, and systems critical to the effectiveness of U.S.

military forces. Specific activities conducted at Nellis cannot be discussed any further than that."

In an attempt to blunt the claims of the suit, the Air Force allowed Environmental Protection Agency inspectors into the base, but did not release any information about what they had found. It simply promised to abide by the environmental laws.

———

As part of the case materials, Turley obtained a copy of a Groom Lake security manual, and before long Glenn Campbell had posted it on the Internet. The government responded absurdly, by retroactively classifying the document.

The thirty-page booklet, of which there were several copies in multiple revisions, bore on its front cover the words "Det 3 SP Job Knowledge." "Detachment 3 Special Police" was the assumed meaning of the initials.

It appeared to be the security manual for Dreamland and included a list of radio code names, procedures, and even maps of the base and insides of some buildings. The maps showed the Scoot-N-Hide sheds—Is this an official trademark?, I wondered—used for concealing equipment from satellites, and the Quik Kill radars and surface-to-air missiles that had long been rumored. There were radio code words for areas and structures. In keeping with the best military tradition, everything had to be renamed. The test site was "Over the Hill," and Rachel was "North town."

For years, there was talk of high living at Groom Lake, and the manual's maps seemed to confirm the legends of Sam's Place, the long-rumored base casino and bar, as finely carpeted and outfitted, the Lore had it, as any in Las Vegas. The manual also offered some confirmation of the tales of fine food at the base, of grapefruits flown in from Israel, of lobsters and other delicacies, of huge spring water bills. It suggested a fleet of Auroras flying in odd delicacies, tucked in the corner of a cockpit, from the antipodes.

Was it real? The manual was crude and klutzy. It seemed unlikely that the Air Force would have put the words "Liberty and Justice for All" on the badges that appeared on its first page. The tone of the code names was unconvincing. "Dutch Apple" for the headquarters seemed inappropriately imaginative—unless it reflected some kind of inside joke.

Procedures were outlined for moving test articles. When back in the civilian world, the "special police" were instructed to say that they "worked for EG&G at the test site." There was quite detailed and accurate information on the operation of the road sensors, facts known to the outside world.

But just who was the manual written for? For the deputized guards, working for Wackenhut or EG&G or other contractors? It seemed to be written just awkwardly enough to be real. It made me wonder again about the MJ-12 documents, which shared some of the same crude explanatory quality. And if the manual was not real, why then had the government sought to classify it?

The government had never before tried such a thing, and by definition information already public cannot be made secret. Did the impossibility of such an effort suggest it was merely a ruse to make the document seem genuine? If so, why?

———

The case came before federal judge Philip Pro. But Judge Pro had previously found the government not liable for damages to some 216 workers who had been exposed to radiation at the NTS between 1951 and 1981—workers like Joe Bacco—many of whom had been assigned at times to Area 51 itself. Pro seemed to believe in keeping security. All he wanted was a letter from the president of the United States swearing that we needed to keep Dreamland in the dark. And he got it. In September 1995, Bill Clinton signed a statement affirming that to reveal what Turley and his clients wanted to know about Groom Lake "could reasonably be expected" to damage the national security.

The government pressed for the names of the John Doe clients, a request Turley felt sure was meant to intimidate the workers. Then Judge Pro ordered the documents in the case sealed. What that meant became clear all of a sudden. In the summer of 1995, Turley was in Chicago at the bedside of his ill father when he got a call: OSI agents were on their way over to the George Washington University Law Center to seal his office. He immediately called his secretary back in Washington and asked her to alert campus security. He had a vision of the bicycle-mounted campus cops in hand-to-hand combat with OSI commandos infiltrating through the ventilation system. Turley's office was officially sealed, but the files relating to the case had been placed in a safe to which only he and one associate had the combination. It was like an embassy of Dreamland inside the District of Columbia.

In 1995, the case against the EPA was dismissed, and, the following year, so was the case against the Air Force. In the fall of 1997, the Ninth Federal Circuit Court took up Turley's appeal. The following year, the Supreme Court would reject his final appeal.

If the case had come to trial, Turley said, he planned to call the secretary of defense, the secretary of the Air Force, and the president's national security adviser. If none of them was willing to admit to the existence of the base, then he said he would call representatives of the Russian embassy. The Soviets, after all, had photographed the base from their satellites.

Turley had the government caught in a post–Cold War half nelson: "While the United States government refuses to acknowledge the existence of this base to the American public," he was able to argue, "the Russian government recently declassified much of its intelligence information as part of a new openness policy following the fall of the Communist regime and the adoption of democratic process." And since the 1993 signing of the Open Skies treaty, the Russians and all other signatory nations had a right to do so.

In other words, it was legal for foreigners to photograph the base, but not for the Americans who had paid for it.

24. Rave

If sinister forces were manipulating the mass media, they were doing it quite effectively. Area 51 was infiltrating television as stealthily as any Hollywood extraterrestrials had ever invaded the planet. It kept popping up, like the Area 51 video game itself, in the oddest places. And the media had their own strange and mythologizing view of Dreamland.

In Las Vegas, which presented subcultures as casino themes, the Area 51 club opened, its big signs in red-and-white warning stripes and stencil letters visible from the freeway. X-TREME PARTY, ESCAPE REALITY, they read. A few months later the club closed, with appropriate mystery.

A whole range of cable shows dealing with the mysterious came to Rachel and talked to Interceptors. Agent X showed up on MTV! And PsychoSpy could be counted on to give good soundbite. His sentences were ambiguous enough to be useful; he seemed to know where the editing would happen.

There was a familiar pattern to most of the programs: some lights over the Jumbled Hills—Janet aircraft or flares, typically—the standard photos of the base, sometimes an establishing shot of Las Vegas neon, talking heads saying there were "mysterious things flying." The key to all the segments was to leave open the question of whether or not there were real saucers hovering over the area like one of those magnesium flares.

After the image of the Manta appeared in magazines, Steve Douglass was flooded with calls from television and print reporters, and even Hollywood producers. Within a few days, the major networks had paid calls on him, and *Unsolved Mysteries*, the tabloid TV series, dispatched two trucks and camera crews to his quiet Amarillo street.

Robert Stack, the host and erstwhile battery pitchman, didn't like the antigovernment tone he saw creeping in to the piece, so they added a line to the script to the effect that "the Air Force denies the planes are theirs. So the question remains, Whose are they?" It was important for the general format of *Unsolved Mysteries*, as in the others, that the "question should remain." The truth had to stay out there.

In April 1994, ABC-TV, while on a shoot, clumsily bumped into the camou dudes, who stopped the crew and confiscated its film, perhaps irritated by the fact that CNN had shortly before set up a camera atop Freedom Ridge and broadcast views of the base.

By May, the press safaris to Freedom Ridge had become so frequent, the viewing points so crowded, that PsychoSpy described a fistfight between two reporters. It amazed the Interceptors, who remembered when few knew the way up at all. In October, Larry King and entourage descended on Rachel. They set up on the side of the road, with the wrong set of hills in the background. No saucer landing in the desert could have looked stranger than Larry's stage set—desk, chairs, lights, and coffee mugs—glowing amid the trampled sage.

Someone mailed Steve a videotape shot by two Las Vegas cops who had read about the TR3A and headed north to Dreamland. Perhaps inspired by beer, they caught sight of something in the sky that danced wildly on the tape, a sign of a camera held by uncertain hands. Their voices were audible, screaming, "It's the fuckin' Manta! It's the fuckin' Manta!" Steve concluded that the craft was probably a B-1.

———

All of a sudden you could find references to Area 51 everywhere. There were scenery files for the popular Microsoft game Flight Simulator that one could download from the Internet to "fly over" Groom Lake. The Marvel company latched on to Area 51, producing a comic or two, and television had embraced it.

An NBC-TV program called *Dark Skies*, set in the mid-sixties, featured aliens digging an underground base beneath Area 51 and Howard Hughes catching on to their plans. "We've got to get to Dreamland" was the most memorable line. A CD-ROM came out with the old cartoon character Jonny Quest delving first into the mysteries of Roswell and then into Area 51.

The story of Area 51 had long held special appeal to technogeeks. One of the Apple Newton software group, for instance, took an interest in it after a trip to Rachel in 1994. He hid a secret feature in one version of the software: If you knew where to click, you could picture Area 51 on the Newton's map. If the user picked Area 51 from the map, the icons in the date book application took on an alien theme—alien faces, flying saucers, robots, and so on.

Then, in August 1995, as the story goes—and we are strictly amid the Lore here—a cryptographer at the CIA was one of the beta testers for the new program. When he saw Area 51, he went to his bosses, who demanded Apple remove the reference. Management "caved in," the sources say, but the feature was covered over rather than removed and there is yet a trick for retrieving it.

Then the notion of overlapping the Generation X demographic and the UFO one began to swarm in the minds of marketing types—the Gen X files, that was the concept. In the second episode of *The X Files*, the popular show that twists the weirdness of *Twin Peaks* into all sorts of conspiracy lores, Dreamland was transferred from Nevada to Utah, where it became "Ellens Air Force Base." "A mecca for UFO buffs," like Groom Lake, it is omitted from USGS maps, and the hills above Dreamland became tall reeds—equally good hiding places.

In this version of the story, the base is rumored to be "one of six sites" to which the Roswell wreckage was shipped. There the round craft built using alien technology became triangular; the Little A"Le"Inn is transformed into a diner with a fat lady who took UFO snapshots off her back porch. Agents Mulder and Scully see dancing lights and encounter hovering craft. They are menaced by men in black with the requisite sunglasses, and a black helicopter dives at them. There is a reference to "the Aurora project," and dabblings with mental reprogramming. "That's unreal," they conclude, then, "I've never seen anything like that," leaving hanging the suggestion of a causal link between the two statements.

————

The redoubtable engine of American marketing, as simultaneously wondrous and horrific as the military machine, had quickly moved to sell teen alienation back to Gen-X. Soon I noticed alien faces, with the almond eyes and big head, everywhere—alien jewelry, alien T-shirts, alien temporary tattoos in the malls, in the hip shops in the East Village of

New York. The alien face had become a wry nineties equivalent of the seventies-era smiley face.

The image of the big-eyed "gray" alien was set in the early eighties by authors Budd Hopkins and Whitley Strieber, and it superseded earlier images of extraterrestrials. If earlier aliens had represented Communist invaders (*Invasion of the Body Snatchers*) or disease (*Alien*) or been depicted as friendly babylike creatures (*Close Encounters* and E.T.), this new one was a huge fetus or hungry child, with big Keane kid eyes. It was also an echo of Munch's *Scream*—the very face of modern angst.

This alien face had long been familiar, but had never been so graphically standardized before. It summed up a growing American subculture devoted to ideas of abduction and implantation that paralleled a fascination with recovering childhood experiences—commonly those of abuse—via hypnosis. In 1995, Testor released a plastic model kit of a standing gray alien, proof that it had become the iconic image of the extraterrestrial, just as the flying saucer was of the UFO.

The new image of the alien was as much ironic as iconic. It was significant that we had begun calling creatures from other planets "alien" again in the eighties, after the previous decade had popularized "extraterrestrial." Much of the new alien material turned on the puns associated with "alien"—the joke was that the grays figured as immigrants. The face had become as much a graphic cliché, an ethnic cartoon, as Sambo or Uncle Tom. Was America's latest favorite ethnic group from Zeta Reticuli? "Do we call them Astro Americans?" a friend asked.

Tropes of the alien often serve as parables for dealing with issues such as immigration. Note, for instance, the differing degrees of irony evidenced in *Coneheads* ("We are . . . from France," and their nemesis in the film is an INS agent) and *Alien Nation*, where the aliens exhibit the irritating traits of various earthly minority groups: They are ex-slaves with strange music; they threaten to take jobs and resources from Earth natives; they eat bizarre food and score intimidatingly high on math tests. *Men in Black* continued the theme, tossing off jokes about immigrant New York taxi drivers.

My favorite T-shirt on this theme depicts a cliché alien wearing a sombrero and bandoleers, bearing the legend: "We don't want no stinkin green cards."

The alien face's iconism was accomplished when it became subject to manipulation of context and ironic reference. Thus T-shirts picture the

Beatles with alien heads, or the "see no evil, hear no evil, speak no evil" trio rendered in alien faces, and a whole host of alien-face pop artifacts—earrings, Schwa artifacts, Alien Factory skateboard graphics.

The alien theme is strong in music. The band Foo Fighters recorded on its own Roswell Records label, and an album by the group Spacehog is called *Resident Alien*, its cover art bearing an extraterrestrial "green card."

Television could handle the pop-alien theme with equal facility as drama or comedy. "Aliens are all around us," intones the narrator at the beginning of *Third Rock from the Sun*, a sitcom in the tradition of *My Favorite Martian*, *Mork and Mindy*, and *ALF*—a view of the outsider as observer as old as Montesquieu's *Persian Letters*.

In the series *Dark Skies*, the alien invasion turns out to have been so subtle and surreptitious that it has touched every major event of the last fifty years, from the shooting down of the U-2 to Project Mercury to the Kennedy assassination. Conspiracy theorists are alienated by mainstream explanations, of course. Concealment and conspiracy is another theme behind the image—it was the logo of the Big Cover-up, the "Cosmic Watergate." "The government is lying," T-shirts told us.

In time, the alien face came to appear to me as the face of suspicion of government—and the projection, perhaps, of the new generation's alienation. "We are not alone," the slogan that often captions the gray face, may be as much an expression of hope as an assertion of belief. Someone once said, "Aliens are alien because we alienate them." That was ALF, the sitcom alien.

―――

By the spring of 1996, Hollywood was turning the Lore from folktale to fodder for commerce. The Interceptors were at once amused, irked, and perhaps a bit sad, resentful that the Hollywood dream machine was taking over their base, dismayed at a crass mercenary effort to cash in on the Black Mailbox.

The idea behind renaming Nevada Highway 375 the Extraterrestrial Highway was to bring tourists to the area. When the Nevada legislature held hearings on the idea, the only witnesses to appear—and they were in favor—were Joe and Pat Travis, the largest likely economic beneficiaries of the idea, and Ambassador Merlyn Merlin, himself an avowed extraterrestrial.

PsychoSpy took a hard line against the renaming, more, one suspects, out of an instinct to oppose government than for his stated reason that no thought had been given to the consequences of bringing tourists to the area and possibly into contact with the camou dudes. If anything, he felt his own bailiwick was being invaded—he was after all the first to produce a tourist guide, the first to lead groups to the perimeter, the first to pioneer four-wheel drive to the top of Freedom Ridge. Now it was all about selling souvenirs. Yet PsychoSpy himself had set this all in motion when he printed up his first T-shirt bearing the invented Dreamland patch.

The dedication of the ET Highway and the unveiling of the road signs that marked it was a ceremony twice hijacked. The first time was by the producers of the film *Independence Day*, which would dramatically change the Area 51 Lore. Whetting anticipation for the summer '96 release of the movie, its stars agreed to join the ET Highway dedication, and the producers donated a "time capsule" to Rachel. This guaranteed that the politicians would be overshadowed.

In front of the Little A"Le"Inn, actors Bill Pullman and Jeff Goldblum moved among a thin crowd and posed in front of the signs as Nevada tourism officials explained that prospective visitors could call an 800 number for an "ET Highway Experience" package complete with map. The governor joked that perhaps the signs should have been placed so they could be read from above.

A well-known state legislator named Bob Price, an eccentric and colorful character who led "fact-finding trips" to the cathouses, appeared in Darth Vader costume. "You're Bob Price," shrewdly commented a Rachel youngster, looking right at him.

"The only aliens I've seen are the people who visit here," a little girl told Mary Manning, the reporter for the Las Vegas *Review-Journal*, and this youngster was more correct than she knew.

The event was hijacked a second time on the highway itself. While PsychoSpy boycotted the dedication, the Minister and Agent X rode along in the convoy, which began in a parking lot in Las Vegas and headed up to Rachel for the unveiling of the official ET Highway signs along Highway 375. They portrayed the silhouettes of flying saucers and—no ET here—an F-117 in silhouette.

Agent X led the way in a rented red LeBaron; the Minister's CRX was in fifth place. There were about thirty cars and a big charter bus. As they came down from Hancock Summit into the Tikaboo Valley about thirty miles south of Rachel, just at the point where the Groom Road stretched out to the west, looking as always like a pole of dust rising straight into the air, the Minister caught sight of a bright yellow sign stuck into the dirt by the roadside, with an arrow to the left and the official ET Highway logo. Soon the whole convoy was rumbling in a cloud of dust down the dirt road, straight toward the Area 51 perimeter.

It was a plot by the Interceptors, code-named Operation Coyote, after the cartoon character Wile E. Coyote, who is constantly posting fake road signs to divert the Roadrunner.

The Minister decided to pull off before he got to the guard shack he knew lay a few miles ahead. He understood the rule: You're under arrest once you get to the shack, which is on the wrong side of the perimeter.

Then a Nevada highway patrolman realized what was happening and came roaring up, siren wailing, lights blazing. Through the dust ahead, the Minister could see the lead car taking a sharp right-hand turn onto a dusty road that doubled back toward route 375, through the Medlin ranch. But the planners of the diversion had clearly hoped that it might go all the way—the governor of Nevada and other dignitaries, the whole motley movie and business crowd arriving at the perimeter. Hell, at the guardhouse!

———

Independence Day, which set box-office records by grossing nearly $150 million in its first two weeks of release in July 1996, established Dreamland in the popular mind—but with a twist. The film provided a key new link in the Lore. It tied Area 51 directly to Roswell, whose legend was also growing daily. While, of course, the traditional story had tied Roswell to Hangar 18 at Wright-Pat, the legendary repository of recovered saucers and bodies, *Independence Day*'s story had them ending up at Area 51. At one point, the president says disdainfully, "I can assure you there is no Area 51." "Well, Mr. President," the head of the CIA responds, "that's not . . . exactly . . . true."

Area 51 becomes the headquarters in the movie from which Earth resists invasion. The president asks the questions we were all asking about Area 51. "How come I wasn't told about this place?" "How did they keep this secret?" "How did they pay for it all?" But since Area 51 ends up saving humanity, the implication is that we should be grateful it was there. Thus millions of people heard about Area 51 for the first time.

Hollywood's Area 51 looks more like we'd imagined it than the real one: It is slicker, shinier, more sci-fi. The film conjures up an underground lab with tilted glass walls and aliens stored in giant lava-lamp-like containers. It's packed with high-tech equipment: all the war rooms and secret labs of a dozen films of the past rolled into one. It looks, in fact, something like the Area 51 in the video game.

In the early nineties, Ed McCracken, the CEO of Silicon Graphics, whose workstations are used both to devise new aircraft designs and to produce movie special effects, declared that the demands of mass media had supplanted those of the Pentagon as the engine of technological innovation. A 1996 Air Force report on the future, called "New World Vistas," declared that "entertainment organizations" had the skills and means to produce better simulators than the military.

Was Hollywood supplanting the Pentagon? Would the Dreamworks movie studio be the future source of Dreamland's technology?

———

"Calling all 'Encountered People'!" read the proclamation that appeared on the Internet in August 1996. An outfit calling itself Zzyzx Productions and "The Center for the Study of Aerial Phenomenon" announced "Abduction, live at Area 51. An all-night political action rally and UFO-watching vigil" and "rave party." The fine print coyly declared an intention to "encourage peace, love, and harmony, so leave your ray guns at home." The tickets, twenty-five dollars a pop, would be available through TicketMaster.

That seemed reasonable for a pass to Area 51, only the party turned out to be scheduled for a lot behind the trailers in Rachel. The music would be techno—the robotic dance stuff of the new Germany, steeped in the dust of the Wall, now imported, manipulated, and cut to street strength. The idea seemed to be that the fellows at the base might warm to this New Age Woodstock. What I was hoping for was something more like the Saucerian conventions at Giant Rock.

The road was familiar to me now, but it seemed somehow different—richer—with each trip. The landscape's browns and tans seemed to contain rather than exclude colors, if not outright reds and greens, at least what the red and green brown might dream of being. In the little settlements along the way, a few optimists attempted to fight the browns of the desert by painting their houses and stores in bright aqua or turquoise. It took me a while to see how shrewd a choice that color was, how that turquoise sang out against the landscape. It was the direct opposite and a powerful antidote to the oppressive hue of the desert.

On the way up to Rachel I stopped for gas. As I paid and came back to the car, I noticed a world-weary guy with a sleeveless shirt and a beat-up pickup. "How are you?" I asked.

"A little closer to somewhere," he answered, as if trying to convince himself.

And I almost said, "But still a hell of a long way from anywhere."

When I finally reached the Black Mailbox, I spotted a Camry with Arizona plates parked beside it. I pulled over to talk to a young couple who stood looking off toward the Ridge. The man was a stockbroker. "They say this is the place," he commented, dreamily. "We drove all the way from Tucson, just to see."

―――――――

The most romantic thing about the rave was the dust swirling in the big floodlights. The promoters had promised "sunbaked desert dance dirt," "fire-breathing tribal drum circle," and all-night dancing in the shadows of the Jumbled Hills. They held open the possibility that the boys at the base might be tickled enough to float one up just over the Ridge, offer a hint of the mysteries beyond. The partygoers I talked to knew of Area 51 only as a saucer site. They were ignorant of the history of the U-2 and the Blackbirds.

In Rachel, the locals—piqued by the prospect of drugged and drunken youth from as far away as Los Angeles—watched with interest. The sheriff's office required the promoters to post a large bond, and deputies' cars patrolled the area. Strange vehicles—Woodstock-era Microbuses, junker compacts with out-of-state plates—began to appear in front of the Quik Pik and the Inn.

At the Research Center, some of the Interceptors gathered to watch. In the little yard by the trailer, they dipped chips and roasted hot dogs and marveled at the speed with which the media machine had latched on to the mythology of Area 51.

"It's become the dominant urban folk legend of the nineties," Zero said in the kitchen, unwrapping more chips. Behind the trailer was a little shed with a platform on its roof that turned out to be handy for viewing the preparations. We climbed up to look at the assembling trucks and lights and speakers. It was Little Freedom Ridge, a mini-Tikaboo, but it would bear the weight of only three or four people.

I had checked into a motel up the road in Alamo. The lady at the desk told me they had a special deal: two bucks extra for five channels of TV, five dollars for the cable and ten channels more. I went the whole hog: a better rate per channel. Besides, I felt a need to stay close to the umbilical cord of mainstream culture.

I lay down for a few minutes in the afternoon and in a groggy sleep dreamed that I had figured out the secret of the numbering system for the areas at the test site, which appeared randomly on the map. It all had to do, I dreamed, with an angle of the border of each area from the north-south axis. When I awoke and looked again at the map, I realized the dream scheme made no sense at all, that it was these sorts of angular alignments that the supporters of the Mars face, the believers in ancient civilizations and secret bases on Mars and the moon, used to support their case.

The individual's dreamwork is echoed in that of his culture. The same strategies of compression, substitution, abbreviation, displacement, and symbolism Freud sees in individual dreams may apply, I realized, to the shapes of tales in the Lore.

If you believed that dreams were worth looking at as a way to understand a person's hopes and fears, then wouldn't looking at the dreams of a culture accomplish the same thing? Couldn't the fascinations of its core be written in the obsessions of the fringe? A tunnel at the nuclear test site evolved into a network of underground railroads, perhaps, a MiG was transmuted into an alien ship, a flare into a saucer's light. I thought of the tales of the footprints of deer melting out in the sun into those of the imagined Yeti.

Any dream expresses a wish, Freud states, but how could some of the dark and frightening dreams I had heard be wishes? They saw the source of the fear discharged, was Freud's answer. They granted a wish, too, for order and explanation—dreams crystallized vague fear into a specific bogeyman, which one could better comprehend. Couldn't the fear of a new world order be an expression of a desire for order; couldn't the arrival of aliens save us by organizing us to resist?

The souvenir vendors had arrived first thing in the morning. The latest item was a T-shirt showing a saucer over the lake bed and the legend "Area 51 Yacht Club." One vendor, an enormous man selling glow-in-the-dark alien heads, T-shirts, and charms, told me he used to be with Navy Intelligence. He sat in a minivan beside the Rachel Quik Pik, wearing a SEAL team T-shirt and an LAPD bomb squad cap.

"Naval Intelligence," he repeated. "Ever hear of Richard Marcinko? Seal Team Seven. It's not supposed to exist, but it does."

He had strong opinions on Bob Lazar's story. "That W-2 is as real as can be," he said.

———

That night, huge screens surrounded the circular dance floor, flashing music-video images back on themselves, reminding some of old drive-in movie screens. But only a few dozen dancers showed up, groggy after the long drive dodging the cows that sat on the edges of the ET Highway. The bitter alkaline dust stung the eyes and seeped into every fissure of clothing and body.

A few misguided Hollywood types ended up in town. At one point a limousine turned in to the parking lot and I caught a glimpse of a softly lit interior, packed with cut-crystal decanters glowing like artifacts in an old-fashioned sci-fi film. Then the dust rose up and covered it all.

The UFO souvenirs failed to sell well. At the end of the evening, I caught sight of the Naval Intelligence man still sitting in the minivan. There was no evidence he had ever left it.

25. Remote Viewing; or, "Anomalous Cognition"

At the rave, the promoters had lined up a series of real-life "abductees," who sat at card tables arrayed under tents looking ill at ease. Among them was a woman who did not claim to be an abductee but was willing to talk—a lot, very fast, and in run-on sentences—about black helicopters, Tesla, thought bubbles, interdimensionals, and portals. Her name was Kathleen Ford, and around the time I first climbed the Ridge and looked down on the base, she'd begun taking pictures of strange floating or flying objects along Mailbox Road, looking west over the Jumbled Hills toward Dreamland.

"At first I wanted to take pictures of UFOs and sell them to magazines and make money," she told me. A blackjack dealer in Las Vegas, she would come up every few weeks and shoot day and night.

Ford was clearly smarting from a long history of encountering skepticism—how often had she heard that this image, say, *couldn't* be a flare, or that one was surely *not* the effects of lens or diaphragm. She pointed out one photo that was shot on Easter Sunday, a holiday that even the denizens of Dreamland respected, she said, and on which they did not fly.

I had seen some of these snapshots on the wall at the Little A"Le"Inn, along with all the other greasy, dusty, spotted images of lights in the sky. They were all carefully labeled with details about the camera and film used. In almost every instance, the name of the camera was misspelled. The captions included as much specificity about the time, date, and equipment as there was a lack of specificity about their content. "Two visible ships taken by Mail Box Road Cannon with 200 Zoom Kodak Gold 200." Or, "Invisible ship with light beam going below mountain. This photo was shot facing west at Mail Box Road at 7:50 A.M. Fugi Automatic with 80 zoom, Kodak T-Max, 400 B/W."

One word of that caption caught my attention: *invisible.* As in: "This invisible object appeared after I experimented with music."

By invisible, I understood her to mean that things showed up in the photographs without having been visible when the shutter was snapped.

"That's when I got the eyeball," she said.

"The what?" I asked.

"The eyeball. I give them all names and this one I just call the eyeball. It's translucent."

Indeed it could be an eyeball, floating in front of the flash-lit, out-of-focus grass by the highway, the soft LED digits of the dating function visible in the lower right-hand corner. Emerson's transcendental eyeball, Jung's eye in the sky—whatever you wanted to call it.

"After I got this one I went, 'Oh . . . my . . . God.' I cried for three weeks. They've lied to us, I thought. When I saw this, everything I had read about UFOs and had dismissed suddenly became feasible and I cried, cried, cried, cried."

————

I picked up the book Ford said had inspired her. The paperback cover of *Silent Invasion,* by Ellen Crystall, Ph.D., bore an image of an alien face, like a film still. Inside were lots of photographs that resembled Ford's, pictures of "Tesla globes," spaceships, even aliens in Westchester County, New York.

Crystall was the clear source of inspiration at least for Ford's captions: The author, with her apt New Age name, had supplied the same details of camera and film type for her photos. Here was a typical Crystall caption: "Large Tesla Field. Taken: June 12, 1988, at Pine Bush, New York. Camera: Nikon 35mm SLR with 50mm lens. Film: Kodacolor negative print film (ASA 400). Exposure: 1/60 sec. at f//1.4 with flash." Elsewhere, she supplied the name of her developer: Fotomat.

Crystall's globes and ships could also have been drops of some kind of staining liquid on the film or lens, but she saw them as Tesla bubbles and beams and ships and aliens. She might not have seen it unless she believed it. There was a twist: While Ford had photographed UFOs she couldn't see with her eye, Crystall claimed to have seen UFOs that didn't appear in her pictures. Some UFOs, she believed, generated shortwave or other radiation that made them invisible. She had seen triangles in Westchester County that resembled the black planes seen in Nevada and California. But realizing that such planes were not generally tested in populated areas, she concluded that "there may be forms of stealth aircraft that are 'true' UFOs built and operated by human beings."

She argues that the B-2 Stealth bomber shown to the public "is really a decoy to divert attention from where the money and effort are really being placed—namely, on construction of enhanced stealth craft capable of hovering at ground level, cruising at speeds ranging from slow walk to thousands of miles per hour, and turning invisible to the human eye. In other words—American UFOs."

Crystall's and Ford's photographs reminded me of the strange and sinister forms, like malign bacteria, seen through a microscope; or of the atomic blasts in Harold Edgerton's photographs—no vision of an alien invasion could ever conjure up a more sinister-looking life-form than these death-forms, slices of vision thinner than the human eye could seize. Ford's photos were just as disturbing.

———

Invisible craft made visible—that was Ford's goal. I was reminded of a chapter in the Air Force "New Vistas" report boldy headed "Invisible Airplanes." The report talks about all the planes of tomorrow, unmanned, invisible to radar, to infrared, to the human eye. The fighting robot planes—UCAVs—that would succeed UAVs would have even stranger shapes. They might look like tailless triangles, like pumpkin seeds, even like discs. "Potentially," said Air Force planners in *Aviation Week*, "a saucer shape could produce the most maneuverable UCAV if combined with vectored thrust for maximum lateral agility. Moreover, a weapon could be pointed by simply rotating the aircraft without altering course."

Stealth makes airplanes almost invisible to radar but not to light, so they fly only in the dark. "We rule the night," Lockheed's ads bragged. But not the day. The next step was to do for vision what stealth did for radar: create high-tech camouflage.

There was pretty good evidence that some of the planes flying in Dreamland were already making themselves invisible by day. They wore electronic skins. They used a technology that was like wrapping the whole airplane in the liquid crystal display of some laptop-computer screen, turning it into a fabric that could be laid on, bent, or built up like tile, or mosaic. It was something called Polyaniline Radar Absorbent composite, "optically transparent" except when charged with a 24-volt current that triggers the camouflage receptors. They read the ambient light—its brightness but also its hue—and are adapted to match. These are chameleon airplanes.

One such program had been around for a while, called Ivy. I.V.? For invisible?

I asked Steve about it: Did this mean that the reason people hadn't been seeing any planes flying above Dreamland recently was because they couldn't be seen? If you didn't see them, must they be invisible? "Have there been," I asked before realizing what I was saying, "any confirmed sightings of these aircraft?"

But sure enough, Agent X had recorded one such "sighting" near Warm Springs. He heard the whistling of a passing airframe first, then it was flying so low he felt the pressure of the air change. But he saw it, too, and he explained how: You observe an invisible airplane by seeing its invisibility—its interruption of the stars in the sky above it, and of the faint glow of Las Vegas behind it.

———

Ford's photographs were confirmed sightings all right, proof that you had to believe to see. When she was taking pictures, Ford explained, "I go into alpha beta."

I looked baffled.

"You know, dream state, alpha beta. And these images are very dreamy."

In other pictures, Ford pointed out a different sort of blob. "These here may be interdimensional entities," she explained. They might, in other words, signify creatures from other realms of space-time.

"People told me there are interdimensionals. Aliens that can move in and out. John Lear talked about the government having EBEs—'extra-biological entities'—and the government is having a hard time with them. They keep them in an electromagnetic field but they just drift in and out."

At the Inn, Chuck Clark talked about interdimensionals, too. Far out as it sounded, the idea struck me as one of the most provocative areas of UFO-related thinking. What we had once taken for aliens from another star system, this theory went, might instead be time travelers or visitors from parallel universes. These concepts, it seemed to me, were more worthy of consideration, at least on intellectual terms, than the saucers themselves.

Increasingly physicists, both popular and academic, were writing and talking about such ideas, born of the paradoxes of quantum theory. Aspects of quantum theory seemed to require the postulation of parallel or multiple universes. Space-time "wormholes" made time travel a theoretical possibility. String theory projected scenarios wherein an original universe of twenty-one or thirty-four dimensions might have collapsed into the present four.

According to quantum theory, subatomic particles could apparently be in several places at once. From those bits of quantum doubt, theories of parallel, alternate universes had arisen, like conspiracy theories from Eisenhower's toothache. An alternate universe might be identical to this one, except that I have brown instead of blue eyes. Or, more to the point, a subatomic particle that is *here* in one universe might be *there* in its neighbor.

Quite respectable efforts to solve the quantum uncertainty principle had resulted in scientific experiments postulating a theory of parallel universes. In the late fifties, the respected physicist Bryce DeWitt proposed such a solution. It wouldn't take so many universes, he had calculated, only about ten to the hundredth power. Soon physicists were using the term "multiverse" for the totality of possibilities.

All the little bits of quantum uncertainty, all that black matter, made the universe a kind of sponge of uncertainty. Just as Heisenberg's uncertainty principle and Schrödinger's cat helped translate hard-core science into vaguely understood popular lore ratifying wider feelings about how uncertain knowledge had become, the physics of the multiverse became attached to popular ideas of parallel modes of existence—"other dimensions," in the crudest vocabulary of Hollywood, or interdimensionals.

The youfers were always talking about how the recognition of other intelligences in the universe would produce a change in thinking, such as the one brought about by the discovery that the sun was at the center of the solar system and not the earth, or the European discovery of America. I could understand this: Nobody wants to be part of the crowd jeering Galileo, nobody wants to be the last flat-earther. This was, after all, the Greatest Story in Human History.

But spacecraft bearing almond-eyed aliens from Zeta Reticuli was far less convincing as a scientific revolution than parallel universe theory. With roots in mainstream physics, its ideas seemed both wondrous and possible. In his modestly titled book *The Fabric of Reality*, the brilliant physicist David Deutsch discussed how all of twentieth-century physics pointed to parallel universes. The facts were there, he argued, only bold imagination was lacking for their acceptance. For me, photographs like Ford's came to suggest this element of imagination: I was as seduced by the images as by the ideas. The odd hovering shapes, the soft spray of flash on desert plants and pavement edges, the mere hint of landscape beyond, were like diagrams of the darker, sketchier implications of the new physics.

Another physicist, Fred Alan Wolf, in *The Dreaming Universe*, believed that parallel universes might be the source of schizophrenia, visions, even dreams. UFO sightings, he noted, seemed to many viewers to possess a dreamlike quality. Could UFOs have an existence that was half in, half out of this universe? This was taking Jung's idea of manifesting archetypes into a more literal plane. It was awfully close to the Borderlands folk or the contactees and their "ether." The things that were seen in the sky, this new way of thinking went, might inhabit some realm halfway between the state of being a thought and the state of its material existence. The key question, Wolf said, was how matter gave rise to thought, "How does meat dream?"

Are these visions created by psychic disturbances? How literally are we to take the idea of objects in the sky being manifestations of cultural unease? The idea of one universe as the dream of another gave Dreamland a whole new meaning.

————

"Now, this one I took on the border," Ford told me.

It showed one of the familiar round metal sensors, those strange mirrored spheres, that mark the perimeter of Dreamland. But there was something else in one corner. "See this?" she said, pointing to what looked like a boulder or a blob. "I think this is a remote viewing blob." The camou dudes, she suspected, could carry out remote viewing of the border from their guardhouses. I did not ask why they bothered to head out in Jeeps and Blackhawks if they could do this.

But remote viewing at "the remote location" seemed eminently appropriate. Hadn't the Army taken the technique seriously enough to spend tax dollars on it?

Remote viewing—the ability to see at a distance—is a paranormal technique on which the CIA and the Pentagon had spent about $20 million and twenty years. Also known by the wonderful phrase "anomalous cognition," the idea was developed by Dr. Harold Puthoff and Russell Targ at the Stanford Research Institute in the 1970s. Their first and prime viewer was an artist named Ingo Swann (Hollywood could never come up with names like these), who directed the effort to turn remote viewing into a useful military intelligence tool. Fearing an emerging "psi spy" gap with the Soviets, the CIA began funding remote viewing and, later, handed the research off to the Army.

It began with some degree of scientific rigor, with the finding that some people did a better job of, say, picking cards facedown on a table several rooms away than they should have by chance alone. "Psi-hitting," it was called. Then in Project Grille Flame, later Scannate and Stargate, it was applied to such tasks as discovering the locations of Soviet submarines or finding hostages in the Middle East. But results kept turning up that embarrassed the Army. When the viewers were directed to search for secret Soviet aircraft, they came back with reports of UFOs.

Viewers were sometimes led through brainwave feedback and other techniques in order to become more sensitive receptors. Two of the early remote viewers were Ed Dames and Joseph McMoneagle. Dames claimed that "we employed people who used altered states to take a look at the radio station in Tehran, Iran, prior to our aborted rescue attempt." Strategic locations in Iraq were another target. The lack of success of those efforts makes one skeptical about remote viewing.

I tried to think of remote viewing, perhaps charitably, as equivalent to frustrated police turning to a medium to locate a body. The program was operated in a series of shedlike buildings at Fort Meade, in Maryland. After the military program ended, remote viewing moved into the private sector. Ed Dames established a firm called Psi-Tech that did "business research," or, less politely, industrial espionage. For an auto company client, for instance, his viewers "go into this library in the sky, if you will, what we call the matrix, the collective unconscious, [and] pull out designs that were Japanese and German."

Other alumni of the program were less positive. Joseph McMoneagle disparaged Dames. Another veteran, David Morehouse, wrote a book titled Psychic Warrior (1996), in which he declared that the feds had recruited him as a remote viewer and then made his life miserable.

An associate professor of political science at Emory University named Courtney Brown, whom Ed Dames had taught RV, had established an outfit he called the Far Sight Foundation, and claimed to be able to view inside the Oval Office and to visit secret bases on the moon and Mars. He envisioned our Mars probe being destroyed by a defending alien craft. When he appeared on Art Bell's *Coast to Coast* radio show, he suggested that the comet Hale-Bopp provided cover for an extraterrestrial spaceship that was heading for our planet.

The Heaven's Gate cult latched on to Brown's idea and stuck out their figurative thumbs to hitch a ride. Before their mass suicide, they visited Las Vegas and played the slots; some members may have attended a conference on Area 51.

Ford's notion that the camou dudes could remote-view, however, was a new one on me. So was the idea that the presence of these remote viewers might take the form of glowing balls.

"Does it really work?" I asked her.

"Sure, I've been taught how to do it. First, you have to give yourself permission to let yourself invent. And when you understand it's okay to make it up, then they start to appear and you say to yourself, 'Hey, I didn't make *that* up.' "

"Can you remotely look over the hills *here* and see what's on the other side?" I ventured gingerly—over *there*, into the base at Groom Lake, into Area 51, into Dreamland.

"Sure," she said, as if it were the most natural thing in the world. "I've been there. It's empty."

26. The White Mailbox

Maybe Kathleen Ford was right, maybe the place was "empty" (whatever she meant by that). Maybe the secret warriors had folded their tents under cover of night and crept away. Maybe the cuts of the post–Cold War years had reduced the role of the base. Maybe the glare of publicity had made operations untenable.

I remembered the statements a congressman made at the time Whitesides Mountain was annexed to the restricted area. The watchers, he said, were a tremendous inconvenience to the men at the base. They had to shut things down when the watchers appeared. "It's not fair," he reported, almost petulantly. Bill Sweetman thought that the cost of doing business in Dreamland had priced it out of the market, that in the new, austere Pentagon, all the security and expense of moving things in and out was too much. He thought the projects had moved elsewhere.[1]

In another sense, of course, it had always been empty, and that was its attraction. We needed it empty to function as a container for speculation. You could fill it up with whatever you wanted. Or maybe we had all emptied it, squeezed out every bit of speculation, overtaxed that humble collection of Butler metal buildings and big hangars and military-issue dorms, demanded too much meaning from it. Perhaps Dreamland was full now and could hold no more of our speculations or fantasies.

On the Internet, you could find this sentiment: "I hope we never find out what's in there," a buff wrote rather wistfully. "I'd just like to observe something about us Area 51 freaks. As much as we talk about wanting to know what goes on in there, I think that's all just posturing. What would happen if the U.S. government opened its doors to us and let us see all that was going on? Depending on what is there, we'd be either vindicated or disappointed, but we would also rapidly lose interest. What would we focus our attentions on? Where would we go next? . . . The greatest thing about Area 51 is its mystery, otherwise nobody would care."

———

To push suspicion to the limit, some speculated that Groom had long been a kind of Potemkin village, designed to draw attention away from somewhere else, to hold down the armies of watchers the way the plywood tanks and fake maneuvers of Operation Fortitude held down Panzer divisions before D-day.

Maybe the real projects were going on at some long-rumored "new Groom," or "baby Groom," in Utah, in New Mexico, in Alaska, in Australia. "The new Groom" became nearly as fabled among the stealth watchers as the original, or as El Dorado among the conquistadores. Was it at Eielson Air Force Base, in Alaska, where Agent X kept his eye out but whose vastness made Groom look like a golf course? Or Pine Gap in Australia, perhaps—rumor had it that several Northrop aerodynamicists, including the legendary John Cashen, had moved to Australia. To Utah, near Dugway and the dreaded storage area for chemical and germ warfare weapons? One top aviation journalist, who told me everything had been moved from Groom, said, "We've heard the pulser jet in the Southeast, out in the swamps." Or was it all moved, as Steve Douglass had heard, to a new secret base over the hill from White Sands in New Mexico? Steve Douglass and I went to look.

———

I had driven across the lava plains north of the White Sands Missile Range, a few miles north of the Trinity site, where the first atomic bomb was detonated. I had passed the northern entrance to the range, called Stallion Gate, when a white Blazer with official plates fell in behind me. The driver was speaking into a mike, its corkscrew of cord trailing behind. I imagined him checking up on me and grew nervous for no reason.

I passed through the Valley of Fire, a landscape of rough stones that resembled coral, as if a whole beach of lava had been laid bare by a receding tide. There was a rolling quality to the depressions and outcroppings, and you could almost imagine that the rock was still liquid. Driving across it, I could understand how looking at the relentless distance day after day could inspire despair—the despair early settlers felt and tried to treat with whiskey and patent medicines.

Steve and I parked across from the fence at Holloman Air Force Base, where the Aquarius briefings said the saucers had landed for treaty negotiations, and Stealth fighters transferred from Tonopah trained in daylight now. Traffic whizzed by with a heavy rush. Binoculars offered a terrorist's-eye view of a base that was like a movie set of an airbase: tower, water tank, palm trees. The F-117s kept taking off over our heads, along with black T-38s. The shadows slid across the pavement, which itself shimmered in cheap mirages.

The next day we stopped by the local BLM office, located in a modern sandstone structure trying to look like WPA Moderne and failing. Three empty government-issue office chairs held a conference in the lobby. I noticed that they were the same kind of chairs as in the photographs of the Roswell wreckage, in Gen. Roger Ramey's office. I suspected I was overconnecting again.

We looked through the big maps, flipping page after page until we found the right ones. Steve focused on the valley west of the mountains that sheltered Holloman and the space harbor.

We stopped at a Dairy Queen to study the maps. A Mexican man with a black Mephistophelian beard but contradictorily patient and gentle eyes walked in. On his shoulder was a tattoo unlike any I had ever seen. I tried not to stare at the tattoo, but it was irresistible. It showed a shapely woman wearing nothing but a gauzy blouse and bandoliers of cartridges. The more I looked, the more the image seemed to deepen and become solid. It shimmered like a printed reproduction of a photo—stand far enough back and the dots merge, the image comes to life. Depth establishes itself behind surface, signal overwhelms noise. I wanted the message of the maps to become that clear, to tell us openly whether there was a new base, and where behind the mountains it was hidden.

We focused our search on the Oscura Mountains. There was a new restricted airspace, number R5107, and we first studied the aeronautical charts, the spaces marked mostly purple and brown, then looked at the more variegated palette of the BLM maps, indicating the usage and ownership of land with its melons, blues, and yellows.

We trudged across White Sands, aiming at the tower of the old Northrup Strip. Now they called the old strip "Space Harbor." All we could see were the top of its antennas and water tower, which barely peeked above the white dunes in a thousand advertisements, when it had stood in for the Sahara, and for Mars.

Steve had been here before at night. Creeping over the brow of the last dune, whiter than white by moonlight, he had seen the base unfold, crisscrossed by huge laser beams and dotted with multicolored lights. Word had it they'd put in the most powerful runway lighting system on the planet. The shuttle astronauts could see it from space. They'd landed here once when bad weather spoiled the usual landing strips at Edwards or the Cape, and for the first time the TV crews were kept away from the landing.

Now in the daytime we crawled across the sand. The more we looked at the maps, the more we drove and wandered through the rippled dunes, the more hopeless and foolish we felt. You would have to have up-to-date satellite photos, an airplane, and free access to the airspace to have even a prayer of finding anything.

———

Yet there was always a Dreamland somewhere down the road. Its very name kept turning up in the oddest places. I learned of a barbecue place in Alabama and a spiritualist outfit in California called Dreamland. One Area 51 buff recorded with excitement, "So, I'm driving back from Costco, listening to the rockabilly show on KCMU and, I am not making this up, this song comes on about a rockabilly cat who meets up with a space alien." The alien asks to be taken to some place called "the Dreamland Bar and Grill."

———

"This New World Order is quite fucking real," Joe Travis said from behind the bar. A few minutes later he launched into an informal karaoke version of "Little Red Riding Hood," by Sam the Sham and the Pharaohs, which had come on the radio. It lent a nice air of menace to his warning. But Joe's act was wearing thin. There was a new mood around the Inn. Tourism in Rachel had become a tired joke—"Area 51" was the punchline—and some of the Interceptors were becoming embarrassed by the whole thing. The Minister had had enough of the Interceptor gatherings, and Mahood issued a "final report" decrying Lazar as a liar and went off to graduate school to study physics.

Campbell, for his part, was growing both more distant from and more possessive of the place. When a magazine report, riddled with errors, charged that Area 51 had been shut down and its activities moved elsewhere, he reacted not just with derision but with something resembling personal affront. He saw himself as webmaster and moderator now, and something like superintendent, too.

When it seemed interest might be waning and new material about the base growing scarce, he even manifested an interest in stories of black aircraft that he had previously shunned. He recounted, with uncharacteristic credulity, a tale of seeing aircraft land on the base and then disappear, as if taxiing underground.

At the same time, he was tiring. In the summer of 1997, he married Sharon Singer, his former assistant in the Rachel Research Center, and began to spend more time in Las Vegas with her and her children.

Much as they might denounce the abuses of government secrecy, the watchers had been drawn there by the mystery, and it seemed to me mystery was a thing in short supply in the contemporary world. That was why so many TV shows and movies worked so hard to provide it. Just as wilderness feeds and nurtures a society that is overcivilized, mystery nurtures a society that is overinformed. The unknown and unpredictable were rarer and rarer qualities in a world of vast information storage and retrieval systems, of sophisticated planning, scheduling, and prediction. We had a fundamental need for uncertainty (as much as we do for order), but not necessarily the kind government secrecy provided. In the spring of 1997, a report from a congressional committee that brought together such odd bedfellows as Jesse Helms, Pat Moynihan, and Lee Hamilton proposed declassifying anything older than ten years, with, of course, the usual "special exceptions." The committee estimated that there were some one and a half billion pages of classified documents more than a quarter of a century old. It was a huge time capsule, requiring expensive maintenance.

The usual talk of means of "penetrating" the perimeter continued: a model airplane, a balloon, even a radio-controlled model car. Norio Hayakawa had his own scheme. He began talking of a "Million Man March" to the perimeter on June 6, 1998, the date he had said provided a conglomeration of multiple sixes, when something dark and dangerous would happen.

One Saturday in late April 1997, four SUVs pulled out of Rachel and drove north on Highway 375. About twenty miles north of town, they turned left on a gravel road, rambled toward Dreamland, and pulled up to the new perimeter line of the restricted area.

Some twenty kids emerged and began setting up easels and canvases in a neat line, about six feet apart and at a 45-degree angle to the vista. Led by a man named Joel Slayton, who taught at San Jose State University, these young art students were collaborating on what Slayton called "a site-specific conceptual artwork involving landscape painting as counter-surveillance of Area 51."

Painting the landscape in old-fashioned oils and acrylics, they were members of "the CADRE institute" (Computers in Art and Design/ Research and Education), who thought deep thoughts about the nature of art and information and how computers figured in to it.

As camou dudes trained their binoculars, Slayton felt a little creepy. What the dudes made of it all, one can only imagine.

The group hauled their finished paintings back to Alamo, where they drew a curious crowd at the local gas station, just around the corner from the first ET Highway sign.

Slayton's official manifesto declared, "The social banality of landscape painting and painters was strategized to be used as a means of countersurveillance by the surveyed, serving as a no-threat typology of threat. In this context the artists demonstrate a perception of art as safe and innocuous, permissible and lacking in relevant information content. The need to surveil such activity is both necessary and unnecessary simultaneously." Now Dreamland drew artists, who drew it.

The camou dudes, Slayton thought, were "serving as a critical agent to assess the significance of the event and resulting information liability." The whole exercise, he proclaimed, constituted "critical discourse on the nature of information culture and information systems." It also seemed a pretty good parody of us serious watchers of the area.

Slayton kept calling the place "a simulacrum." No longer a real place, I understood, but "a reality constituted from media folklore, super secrecy, and the government's denial of its very existence." It existed "only as pure simulation, constructed from the voluminous decentralized and publicly assessable [sic] information that surrounds what might be there." It was a "composited identity formed of electronic networks, e-mail correspondence, and media folklore. Area 51's notoriety as a physical and virtual tourist attraction provides a cultural experience as information simulation ripe with conspiracy theory, Hollywood-style potentialities, and the guarantee of being surveilled."

At PsychoSpy's Research Center, the paintings were placed on sale for $51.51 each with a 51 percent commission going to the Center. Buyers were asked to document the location in which each painting would be hung and to "engage in dialogue" about the whole experience via e-mail.

One frequent visitor to the perimeter happened to see the group and grew suspicious. He was sure they were some sort of security force. They had short hair, he noted, and looked like camou dudes. If they were painters, he said, then he was a B-1 pilot.

———

CADRE's project made my line of inquiry seem positively casual and un-pretentious. Its members weren't interested in aircraft or saucers or holographic experiments; they were interested in philosophical "dia-logue." They were among the most abstract visitors yet to the perimeter, highbrow, high-thinking but, from my perspective, jesters still. If I had once naïvely thought that by identifying the physical craft in the airspace of Dreamland we could then solve its riddles, satisfy the conspiratorial and the curious, I now understood that no rational explanations would satisfy CADRE.

Still, the military soon began to take CADRE seriously as a threat. While some observers on the perimeter were sure that CADRE's painters were cleverly disguised government agents, some government agents ap-parently suspected that they were spies or infiltrators. It happened like this: One of the CADRE members had inquired of the Nellis base histo-rian—who had shown me big, locked file cabinets and had so little to offer about UAVs—if Nellis had received any e-mails inquiring about Area 51. This was purely an exercise, since he assumed none would be released. Not long afterward, he was sent, by anonymous e-mail, a list of e-mail addresses at Nellis. Whether this was a prank or a piece of mis-chief remained unclear, but Nellis authorities were not amused when CADRE forwarded reports of its activity to those on the list. Nellis had been spammed.

In June, a few months after an April "paint-in," Slayton saw a van trailing him. The FBI looked into his activities, and the IRS suddenly manifested an interest in the finances of CADRE.

Then the group sent a party to the land, about forty miles away, owned by Michael Heizer, the artist who lived on a huge tract of land near *Complex One,* his largest work to date. When several CADRE members, young enough to view Heizer as a legend, tried to enter his compound and pay a visit, he proved highly unappreciative. Slayton told me that Heizer threatened to charge them with trespassing. So Heizer's place, I hazarded, had become a kind of little Area 51? Exactly, Slayton said.

———

Incorporating Dreamland into a high-concept work of art, as Heizer had, made me speculate again about just how artful were Area 51's own deceptions. Consider that faceted camouflage, the essential form of military deception, of visual disinformation, had been born in art. During World War I, Picasso and Braque stood watching tanks and other camouflaged vehicles roll through the streets of Paris. "Look," Picasso said, "we are the ones who did that."

The principles at work in this most basic form of deception were the same as those of secrecy. Camouflage, like Cubism, offered bits and pieces, shards and facets. Multiple viewpoints, multiple possibilities— that was all that was needed to create noise, to disguise the real signal. Breaking up the shape into parts was the equivalent of compartmentalization, the most valued intelligence strategy.

———

When I had begun spending more time on the Net tracking stealth chasers and youfers, one day, on impulse, I did a search and typed in one word: "dreamland." The Internet, I knew, was well dotted with UFO and black-plane links and sites, but only one reference came back: to Edgar Allan Poe's poem "Dreamland," stashed away in a university collection of great works of literature.

> By a route obscure and lonely,
> Haunted by ill angels only,
> Where an Eidolon, named Night,
> On a black throne reigns upright,
> I have reached these lands but newly
> From an ultimate dim Thule—
> From a wild weird clime that lieth, sublime,
> Out of Space—out of Time.

Bottomless vales and boundless floods,
And chasms, and caves, and Titan woods,
With forms that no man can discover
. . .

For the spirit that walks in shadow
'Tis—oh, 'tis an Eldorado!
But the traveller, travelling through it,
May not—dare not openly view it;
Never its mysteries are exposed
To the weak human eyes unclosed;
So wills its King, who hath forbid
The uplifting of the fringed lid;
And thus the sad Soul that here passes
Beholds it but through darkened glasses.
. . .

Poe is the patron poet of Dreamland. In *The Power of Blackness,* my old professor Harry Levin had written, "Poe seemed at home only in Dreamland." He dreamed, another critic has written, quoting a famous phrase of the poet's, of "a happier star." Poe is considered among the "Southern Gothic" writers, those authors W. J. Cash described as "romantics of the appalling." Romantic and appalling—which describes what has happened in Dreamland very well.

Read just right, squinting under the Nevada sun, the poem anticipated Nevada's own Dreamland. "Eldorado" turned into a big old Caddy like those parked at the cathouses west of the restricted area, and "the weak human eyes unclosed" or "darkened glasses" evokes long camera lenses or night-vision devices. "Haunted by ill angels," well, there was the U-2, Kelly's Angel, and whatever other strange winged objects you wished to invoke. "Out of Space—out of Time" recalled Lazar's description of the saucer propulsion system, stretching the space-time continuum, warping gravity, like a hammock's net. The "black throne" stood, of course, for the rule of the black budget.

There was the dry lake itself, I began to fantasize, in the poet's "Lakes that thus outspread / Their lone waters, lone and dead." Warming now to the job like a conspiracist making connections, I latched on to his "fringed lid." A playful look at the security lid, to be sure, and the "fringe" groups who visited there.

There were even stories that the Poe poem had been the inspiration for the control tower name—suggesting that sitting out in an isolated base leads to more reading than might otherwise be expected of military types.

Thinking about Poe carried me back to Freedom Ridge, and what I once saw flying in the airspace of Dreamland: ravens, Poe's totemic bird. My mind then leapt to the raven I had seen in another place, which was closed off but visible, another black box: Poe's own room at the University of Virginia, in Charlottesville.

Young Poe had lived there in 1826, before he was expelled from the university for failing to pay his gambling debts. The wooden door has been replaced with glass, as in a bank or department store. Visitors push a button and a dim light comes on. You can see a crude rope bed with Jacquard coverlet, a desk, a pen, and—some historical license—a stuffed raven. A black bird in an almost black room.

The preservation of such a room as a viewable but unreachable space, part memorial, part exhibit, strikes me as very like the Groom Box. It was like the black world itself—a special exception, a dark chamber in the white and stately colonnade of American life and polity. Thinking about Poe's room, I believe I better understood where the dark visions of the black world fit into the ideal of American order. In the secret vaults in the capital where SAR programs are reviewed, a heart of darkness behind the bright classical façade. In the Black Mailbox itself.

———

One summer day in 1996, I headed back up the road toward Rachel, catching a glimpse from Hancock Summit of the hazy, hovering white stick of road that led to the base. As the road curled around and began its long subtle dip—the Mailbox Road stretch—I settled back into the familiar unfolding of the landscape, the Joshua trees, the range of Jumbled Hills to the west. Coming down the big dip, I nearly drove off the road as something caught my eye: The Black Mailbox was white!

No longer the standard arched rural route job approved by the U.S. Postal Service, it was now a big box of heavy steel, whose door swung on two heavy hinges, with a grab handle from a workshop cabinet and locked with a bright brass padlock. Steve Medlin had stenciled on it his name and route in black.

I walked all around it and noticed that someone had stenciled a tiny black skunk on its back end—a wry comment, perhaps, that this thing was built like the Skunk Works would build it. But it was white now, white as the camou dudes' Jeeps, white as Darkstar, white as the celebrated whale.

At the Little A"Le"Inn, I asked about it. "He got tired of people shooting at it," Joe Travis said of Medlin. "Shooting up his mail and all. Made a new one out of quarter-inch steel plate. Now it would take a thirty-ought-six." He snorted a little laugh.

The steel might resist, but the white paint couldn't. Soon after it went up, someone spray-painted the new box black. Medlin repainted it white. I got the idea this might go back and forth for a while.

There was a black mailbox out in front of the Inn now, but Joe said it was just a replica. I asked what had happened to the original. A man on the stool beside me said that it had been sent to be auctioned off a while ago to raise money for town recreation, but a producer from Hollywood had preempted the sale with an offer of fifteen hundred bucks. This seemed appropriate, but as with so much in Dreamland, it proved impossible to determine conclusively.

Acknowledgments

Many people helped along the way, sometimes in a manner appropriate to Dreamland—without being conscious of it. Steve Douglass and Stuart Brown were vital as sources, inspirations, and friends. Paul McGinnis deserves special mention for help and patience in teaching me all sorts of things. Glenn Campbell deserves commendation, not just here, for his help, but from the public, for his advocacy. The late Ben Rich of the Skunk Works was articulate and honest.

I owe debts of instruction and direction to: John Andrews, Michael Antonoff, Eric Baker, Jim Bakos, Wally Bison, Peter Black, Dale Brown, Lowell Cunningham, R. C. "Chappy" Czapiewski, Mike Dornheim, Mark Farmer, Bob Gilliland, Peter Goin, Joshua Good, Jim Goodall, Norio Hayakawa, Steve Heller, Steve Hofer, Gene Huff, Dean Kanipe, Jon Katz, Frank Kuznik, John Lear, Preston Lerner, Tom Mahood, Mary Manning, Dave Menard, Peter Merlin, Randy Rothenberg, Barry Sonnenfeld, Bill Sweetman, Jonathan Turley, Tim Weiner.

John Pike and Steve Aftergood at the Federation of American Scientists, Derek Scammell at the Nevada Nuclear Test Site, Matthew Coolidge at the Center for Land Use Interpretation have all been helpful in this and many other projects. Special appreciation to Randy Harrison at Boeing, Doug Fouquet at General Atomics, Jim Ragsdale at Lockheed Martin, the estimable Drs. Young and Puffer at the Edwards Air Force Base Flight Test Center history office, and Sgt. James Brooks at the Nellis Air Force Base public affairs office.

For support in work whose subject matter abutted and whose investigations abetted this project: Kevin Kelly, John Battelle, Amy Howarth, John Plunkett, and Louis Rosetto at *Wired*; Anita Leclerc at *Esquire*; Connie Rosenblum and Fletcher Roberts at *The New York Times*; Katie Calhoun, Richard Snow, and Fred Allen at *American Heritage*; Chee Pearlman at *ID*; and Richard Story at *Vogue*.

Thanks to Tom for a vital clip on military monitoring and to Ben for a vital tip on the New World Order. To Steve Guanarccia: I really am going to return your copy of *In Advance of the Landing*, soon and gratefully.

Thanks to excellent book editors along the way: Walt Bode, Bill Strachan, Trevor Dolby, but especially to David Rosenthal, for his vision and confidence, Ruth Fecych, for her care and patience, and the eagle-eyed Benjamin Dreyer and Evan Stone. I am grateful for years of help and advice from my agent Melanie Jackson.

My most important debts are to Kathy, whose support went far beyond her excellent reading and editing, and to Caroline and Andrew.

There were a number of people, of course, whose requests not to be mentioned by name will be honored here, but who cannot escape being appreciated, and thousands of others from whom I learned much through postings and comments on-line.

Notes

CHAPTER 2: THE BLACK MAILBOX

1. Clouds in the desert take on a fascinating variety of shapes. But especially remarked on in Nevada are the lenticular or lens-shaped clouds—clouds that with the scalelessness lent by desert distance can seem very much like flying saucers. Many servicemen in the area, especially those from the East, are struck by them; they send photos back to their relatives and to small-town newspapers. The images are sometimes printed as saucer photos.

   Fascination with lenticular clouds has taken other twists. Youfers looking at paintings and engravings have seen what most people would consider clouds as stylized flying saucers. I picked up a volume of UFO lore that contained a sketch of part of Piero della Francesca's famous fresco series in Arezzo. The lenticular clouds pictured in the book seemed to look like flying saucers. But I looked up an image of the same part of the fresco in another volume and found the clouds quite cloudlike in the original.

CHAPTER 4: AURORA

1. Reports of near midair collisions with mysterious aircraft were often picked up by monitors of airlines' radio traffic. The following is typical: "Last night [March 3, 1996] in the early evening, Flight 573 of America West Airlines was making a routine flight from Dallas to Phoenix when it came very close to colliding with a very, very large triangle-shaped craft over New Mexico at approximately thirty thousand feet. The craft, according to my source, was *not* seen by FAA flight controllers, but *was* picked up by NORAD, due to what was described as a doppler shift. The speed and direction of the unknown is not known at this time."

Another overheard tower transmission at Las Vegas's McCarran Airport:

McCarran Tower/Departure: "United 278 please confirm your heading."
United Flight 278: "Well, I wanted to confirm that. Seems like your head-
    ing's gonna take us pretty close to Dreamland."
McCarran Tower/Departure (Aggressive, bordering on hostile): "United
    278, I have no information on a location called Dreamland!!"

CHAPTER 7: VICTORY THROUGH AIRPOWER

1. Using the Defense Mapping Area charts, Paul McGinnis (Trader) assembled a
   list of other restricted military airspaces:

Area Number	Description and Comments
R-2306A	Yuma West, Arizona (Yuma Proving Ground)
R-2306B	Yuma West, Arizona
R-2306E	Yuma West, Arizona
R-2307	Yuma, Arizona (there is also a tethered balloon on 15,000-foot [4,615 meter] cables that carries a radar pointed south, used to detect drug smugglers. This is located near the north-west part of R-2307)
R-2308B	Yuma East, Arizona
R-2501E	Bullion Mountain East, California (Twenty-nine Palms Marine Corps Base)
R-2501N	Bullion Mountain North, California
R-2501S	Bullion Mountain South, California
R-2501W	Bullion Mountain West, California
R-2502E	Fort Irwin, California
R-2502N	Fort Irwin, California (also includes NASA's Goldstone facility)
R-2505	China Lake, California (western part of China Lake Naval Air Warfare Center)
R-2515	Muroc Lake, California (Edwards Air Force Base [AFB])
R-2516	Vandenberg AFB, California
R-2517	Vandenberg AFB, California
R-2519	Point Mugu, California (U.S. Navy-Pacific missile test range)
R-2524	Trona, California (eastern part of China Lake Naval Air Warfare Center. Includes the "highly classified" electronic warfare facility, the Randsburg Wash Test Range, also known as "Sea Site I")
R-2914A	Valparaiso, Florida (Eglin AFB, Air Force Development Test Center [AFDTC])
R-2915A	Eglin AFB, Florida
R-2915B	Eglin AFB, Florida
R-2918	Valparaiso, Florida
R-2919A	Valparaiso, Florida
R-4806W	Las Vegas, Nevada (Nellis AFB)
R-4807A	Tonopah, Nevada
R-4807B	Tonopah, Nevada

R-4808N	Las Vegas, Nevada (R-4808N covers both the Nevada Test Site and the Dreamland "box" around Groom Lake, which is the rectangular region in the northeast of R-4808N)
R-4809	Tonopah, Nevada (R-4809 covers Tonopah Test Range, used for activities such as F-117 fighter testing and Department of Energy programs, such as nuclear rocket testing in the 1960s)
R-5107A	White Sands Missile Range, New Mexico
R-5107B	White Sands Missile Range, New Mexico, also includes Holloman AFB
R-6604	Chincoteague Inlet, Virginia (used by NASA's Wallops Island rocket facility)

2. According to *The Quiet Fire*, a history of the band U2, the group took its name in the late seventies at the fairly casual suggestion of sometime band member Steve Rapid, who told bassist Adam Clayton about the spy plane and punned it with "you too."

3. Published in 1965, *Mission with LeMay* is a central document in the history of Cold War culture.

One of its most fascinating passages is LeMay's effusive comparison of SAC's organization to a B-58 bomber "weapons pod." He did not latch on to the obvious Freudian conclusions with which the thing fairly screamed; instead he compares it to a jack-in-the-box in describing an inspection:

The chief of the ground crew and one of his men are up on the dock, engaged in removing a metal plate from the fuselage of the aircraft. We stand and watch. Off comes the plate, and there is exposed a labyrinth of silver and wire and plastic . . . tiny colored blobs and shreds. That's a meager crumb, a mere sample of the electronic equipment which is stuffed and geared throughout the stiff flesh of the B-58 . . . Something like the business of that old-fashioned jack-in-the-box you had as a child . . . You look up at that plate, and the fuselage aperture, and vaguely you wonder: how are they going to get that snake back in there?

They'll get it back. And every tuft and every peg and every threadlike wire, and every infinitesimal jewel of the complex array will have been tested and found to be functioning, before that slice goes back on the aircraft—with reptiles arranged in designated position, before the plate is locked. The B-58 is crammed with those thousands and thousands of working warming cooling bits of metal and wire and tubing. Every available cubic inch within the body is occupied by such little monsters and treasures.

. . . And in that beautiful devilish pod underneath, the baby of the fuselage—half-size, but still of the same shape and sharpness, clinging as a fierce child against its mother's belly—the B-58 carries all the conventional bomb explosive force of World War II and everything which came before. A single B-58 can do that. It lugs the flame and misery of attacks on London . . . rubble of Coventry and the rubble of Plymouth . . . Blow up or burn up fifty-three per cent of Hamburg's buildings, and sixty per cent of the port installations, and kill fifty thousand people into the bargain. Mutilate and lay waste the Polish cities and the Dutch cities, the Warsaws and the Rotterdams. Shatter and fry Essen and Dortmund add Gelsenkirchen, and every other town in the Ruhr. Shatter the city of

Berlin. Do what the Japanese did to us at Pearl, and what we did to the Japanese at Osaka and Yokohama and Nagoya. And explode Japanese industry with a flash of magnesium, and make the canals boil around bloated bodies of the people. Do Tokyo over again.

The force of these, in a single pod.

One B-58 can load that comprehensive concentrated firepower, and convey it to any place on the globe, and let it sink down, and let it go off, and bruise the stars and planets and satellites listening in.

Every petard, every culverin, every old Long Tom or mortar of a naval ship in the eighteenth or nineteenth centuries, every turret full of smoky cannon at Jutland . . . Big Bertha bombarding Paris . . . musketry of the American Revolutionary battles or the Napoleonic ones. Spotsylvania and Shiloh and the battles for Atlanta. All the paper cartridges torn with the teeth, and all the crude metallic cartridges forced into new hot chambers. . . . Firepower. All the firepower ever heard or experienced upon this earth. All in one bomb, all in one B-58.

He went on: "The B-58 was and is symbolic of SAC . . . If you removed that plate from the body of SAC, you could look in and see people and instruments. They would be as the intricate electronic physiology of an airplane today: each functioning, each trained, each knowing his special part and job—knowing what he must do in his groove and place to keep the body alive, the blood circulating. Every man a coupling or a tube; every organization a rampart of transistors, battery of condensers. All rubbed up, no corrosion. Alert."

The book also includes LeMay's statement that while the Air Force had never intentionally concealed information on UFOs, there were many sightings for which it was never able to satisfactorily account.

CHAPTER 9: IKE'S TOOTHACHE

1. For other variations on the Men in Black theme, see Scott Spencer's novel (Knopf, 1995) of the same name, about a literary novelist whose work-for-hire book on UFOs becomes a runaway bestseller, and *Men in Black* by John Harvey (University of Chicago Press, 1996), which delves into the long and complex semiology of black male attire, from Dracula to drag, Johnny Cash to Johnny Depp—without mentioning "the UFO silencers."

CHAPTER 12: LOW OBSERVABLES

1. The Soviets were so skeptical of the idea of stealth that they ignored the implications of this study. In *The Gulag Archipelago*, Alexander Solzhenitsyn describes how an imprisoned scientist who suggested a stealthlike program was considered insane.

2. This quality of pornographic titillation extended to images of alien spacecraft and alien bodies as well. In the September 1996 issue of *Penthouse*, publisher Bob Guccione ran what can be described as a cheesecake shot of a dead alien—as a centerfold. It was most likely a photo of a prop from the Showtime network's Roswell film.

CHAPTER 19: THE REMOTE LOCATION

1. Steve Douglass was astonished to see the pictures of Tacit Blue. He had seen this plane, he told me, before he'd begun investigating secret aircraft. He and his wife had been on vacation in New Mexico when they caught a glimpse of the thing sailing through a canyon, almost below them and the road on which they were driving. Indeed, it was that sighting that piqued Douglass's interest and inspired him to look into the whole world of secret planes. But at the Air Force Museum ceremony, there was no reference to the testing of Shamu in New Mexico, which probably occurred at the White Sands radar cross-section testing facility.

CHAPTER 21: SPACE ALIENS FROM THE PENTAGON AND OTHER CONSPIRACIES

1. Black aircraft continued to become part of conspiracist mythology. During the 1992 presidential election, accusations were made stating that Vice President George Bush had secretly flown to Europe in the backseat of an SR-71 to meet with Iranian emissaries as part of the Iran-Contra deal. Only by means of an airplane as speedy as the Blackbird, it was believed, could this trip have been concealed.

2. The twisted connections between political conspiracy and saucer lore are further illustrated in *Popular Alienation*, an anthology of conspiracist literature from the Steamshovel Press. In the fevered essays gathered here, aliens mingle with spies and Men in Black, the Moonies, the Trilateralists, and the Bilderbergers, and all connect in seemingly simple, Lego-like attachments of coincidence and suspicion.

3. Trademarks on Area 51 became abundant. One buff tracked trademark filings and found twenty of them, in addition to the video game, trademarked by Atari, covering clothing, toys, and even a registration for the use of the name on "alcoholized lemonade and hard cider, beer, lager, ale, and malt liquor, and carbonated soft drinks, namely root beer."

CHAPTER 22: SEARCHLIGHT

1. Skunk Works buff Andreas Gehrs-Pahl compiled this list of classified program names, the so-called Byeman Code Families, with his speculative interpretations:

- BIG (USAF Reconnaissance projects / missions?)
- BLACK (USAF Intelligence-gathering projects?)
- BLUE (USAF Special electronics missions?)
- BRILLIANT (SDIO/BMDO projects?)
- CHALK (US Navy programs?)
- CLASSIC (US Navy Surveillance / C3I programs?)
- COBRA (Telemetry Intelligence / Surveillance of missile tests?)
- COLD (USAF high-altitude missions?)
- COMBAT (USAF Evaluation of new hardware, test missions?)
- COMMANDO (USAF Special operations?)

- COMPASS (USAF drone/RPV and SIGINT/ECM programs?)
- CONSTANT (USAF development projects / deployments?)
- COPPER (USAF advanced technology studies?)
- CORONET (USAF Electronic surveillance, and deployments?)
- CREEK (USAF deployments?)
- EYE (Weapons developed by Naval Weapons Center, China Lake, CA)
- GIANT (USAF SAC missions and projects?)
- HAVE (DARPA/ARPA or Systems Command [Materiel Command] programs ?)
- IRON (USAF anti-missile programs?)
- OLYMPIC (USAF reconnaissance / surveillance missions?)
- OUTLAW (US Navy Surveillance / C3I programs?)
- PACER (USAF modification and upgrade programs?)
- PAVE ("Precision Avionics Vectoring Equipment")
- PEACE (US DoD Foreign Military Sales [FMS] / MAP programs?)
- PRAIRIE (US Navy Intelligence / SIGINT programs?)
- QUICK (US Army reconnaissance / SIGINT / ECM projects?)
- RETRACT (US Navy programs?)
- RIVET (USAF Electronic intelligence-gathering aircraft?)
- SENIOR (USAF reconnaissance or stealth-related aircraft / systems?)
- SILVER (USAF ELINT missions?)
- TEAL (Surveillance programs?)
- TRACTOR (US Army programs?)
- VOLANT (C-130 Hercules special missions?)

CHAPTER 23: "JOB KNOWLEDGE"

1. Inquiries brought this official response:

> In response to your request for information concerning the Air Force's facility at Groom Lake, Nevada, the 38,400-acre land area once known as "Area 51" was withdrawn from public use by the U.S. Atomic Energy Commission more than 35 years ago under Public Land Order 1662 (filed June 25, 1958).
>
> Since that time, the parcel has been used and administered as a national asset. Because DOE is not now active there, Area 51 no longer appears on maps of DOE's NTS.
>
> Today that land area is used by the Department of Defense as part of its 4,120-square-mile Nellis Air Force Range. For safety and national security reasons, air space above both the Nellis Range and the NTS is closed to commercial aviation and the general public.

CHAPTER 26: THE WHITE MAILBOX

1. Even author Dale Brown dismantled the Dreamland of his fiction: In *Shadows of Steel*, which was published in the summer of 1996, he describes a spy shooting down a super-secret plane at HAWC—his imagined "Hightech Aerospace Weapons Center" at Groom Lake. The incident leads to the closing of the facility and the dispersal of people and equipment.

Bibliography

Assembling these authors and titles, I was struck with a mischievous sense of how the accidents of alphabetization put sworn enemies side by side, pose the conspiracist beside the technologist—a further reminder of how weirdly disparate are the little Dreamlands so many observers have created. It's as if all were lined up—on Freedom Ridge, say—for a group photo.

BOOKS

Steven Aftergood, John Pike, Dorothy Preslar, Tiffany Tyler. *Mystery Aircraft*. Federation of American Scientists, 1992.

George C. Andrews. *Extra-Terrestrials Among Us*. Fate/Llewellyn Publications, 1993.

James Bamford. *The Puzzle Palace: A Report on America's Most Secret Agency*. Houghton Mifflin, 1982.

Timothy Green Beckley. *The UFO Silencers: Mystery of the Men in Black*. Inner Light Publications, 1990.

David Beers. *Blue Sky Dream: A Memoir of America's Fall from Grace*. Doubleday, 1996.

Charles Berlitz and William L. Moore. *The Roswell Incident*. G. P. Putnam's, 1980.

Michael R. Beschloss. *Mayday: Eisenhower, Khrushchev and the U-2 Affair*. Harper & Row, 1986.

Richard M. Bissell, Jr., with Jonathan E. Lewis and Frances T. Pudlo. *Reflections of a Cold Warrior: From Yalta to the Bay of Pigs*. Yale University Press, 1996.

Howard Blum. *Out There: The Government's Secret Quest for Extraterrestrials*. Simon & Schuster, 1990.

Paul Boyer. *By the Bomb's Early Light: American Thought and Culture at the Dawn of the Atomic Age*. Pantheon, 1985. Second edition, University of North Carolina Press, 1994.

Arnold Brophy. *The Air Force.* Gilbert, 1956.

Courtney Brown. *Cosmic Voyage: A Scientific Discovery of Extraterrestrials Visiting Earth.* E. P. Dutton, 1996.

Dale Brown. *Sky Masters.* Donald I. Fine/G. P. Putnam's, 1991.

Dino A. Brugioni. *Eyeball to Eyeball: The Inside Story of the Cuban Missile Crisis.* Random House, 1991.

C.D.B. Bryan. *Close Encounters of the Fourth Kind: Alien Abduction, UFOs, and the Conference at M.I.T.* Alfred A. Knopf, 1995.

Robert Buderi. *The Invention That Changed the World.* Simon & Schuster, 1996.

William Burrows. *Deep Black.* Random House, 1986.

Martin Caidin. *Fork-Tailed Devil: The P-38.* Ballantine, 1971.

Glenn Campbell. *The Area 51 Viewer's Guide.* Self-published, 1993.

Glenn Campbell. *A Short History of Rachel.* Nevada, 1996.

The Center for Land Use Interpretation / Matthew Coolidge. *The Nuclear Test Site: A Guide to America's Nuclear Proving Ground.* The Center for Land Use Interpretation, 1996.

Chuck Clark. *The Area 51 and S-4 Handbook.* Rachel, Nevada. 1995.

William Cooper. *Behold a Pale Horse.* Light Technology Press, 1993.

Paul F. Crickmore. *Lockheed SR-71: The Secret Missions Exposed.* Osprey, 1993.

Ellen Crystall. *Silent Invasion.* St. Martin's, 1991.

Douglas Curran. *In Advance of the Landing: Folk Concepts of Outer Space.* Abbeville Press, 1985.

David Darlington. *Area 51: The Dreamland Chronicles.* Henry Holt, 1997.

Manual DeLanda. *War in the Age of Intelligent Machines.* Zone/MIT Press, 1991.

David Deutsch. *The Fabric of Reality: The Science of Parallel Universes—And Its Implications.* Allen Lane/Penguin, 1997.

Steve Douglass. *The Comprehensive Guide to Military Monitoring.* Universal Electronics, 1993.

Lawrence Fawcett and Barry J. Greenwood. *The UFO Cover-up.* Prentice Hall, 1984.

Paris Flammonde. *The Age of Flying Saucers: Notes on a Projected History of Unidentified Flying Objects.* Hawthorn Books, 1971.

H. Bruce Franklin. *War Stars: The Superweapon and American Imagination.* Oxford University Press, 1988.

Stanton T. Friedman. *Top Secret/Majic.* Marlowe and Company, 1996.

Stanton T. Friedman and Don Berliner. *Crash at Corona: The Definitive Study of the Roswell Incident.* Marlowe and Company, 1992.

John G. Fuller. *The Day We Bombed Utah: America's Most Lethal Secret.* New American Library, 1984.

D. M. Giangreco. *Stealth Fighter Pilot.* Motorbooks International, 1993.

Peter Goin. *Nuclear Landscapes.* Johns Hopkins University Press, 1991.

Timothy Good. *Alien Contact: Top-Secret UFO Files Revealed.* William Morrow, 1993.

Timothy Good. *Above Top Secret: The Worldwide UFO Cover-up.* William Morrow, 1988.

James Goodall. *America's Stealth Fighters and Bombers.* Motorbooks International, 1992.

James Goodall. *SR-71 Blackbird.* Squadron/Signal Publications, 1995.

Richard H. Graham. *SR-71 Revealed: The Inside Story.* Motorbooks International, 1996.

Richard F. Haines, ed. UFO Phenomena and the Behavioral Scientist. Scarecrow Press, 1979.

George Hall. Nellis: The Home of "Red Flag." Osprey, 1988.

Richard P. Hallion. Test Pilots: The Frontiersmen of Flight. Smithsonian Institution Press, 1988.

William F. Hamilton III. Cosmic Top Secret: America's Secret UFO Program. Inner Light Publications, 1991.

Michael Hesemann and Philip Mantle. Beyond Roswell: The Alien Autopsy Film, Area 51, and the U.S. Government Cover-up of UFOs. Marlowe & Company, 1997.

Richard Hofstadter. The Paranoid Style in American Politics. Vintage, 1965.

David Jacobs. The UFO Controversy in America. Indiana University Press, 1975.

Clarence L. "Kelly" Johnson, with Maggie Smith. Kelly: More Than My Share of It All. Smithsonian Institution Press, 1985.

"J. Jones." Stealth Technology: The Art of Black Magic. Aero Books, 1989.

C. G. Jung. Flying Saucers: A Modern Myth of Things Seen in the Sky. Translated by R.F.C. Hull. Princeton University Press, 1964.

Maj. Donald Keyhoe. The Flying Saucers Are Real. Fawcett, 1950.

Maj. Donald Keyhoe. The Flying Saucer Conspiracy. Henry Holt, 1955.

Philip J. Klass. UFOs Explained. Random House, 1974.

Curtis LeMay, with MacKinlay Kantor. Mission with LeMay. Doubleday, 1965.

Tony LeVier. Pilot. Harper & Row, 1954.

Michael Lindemann, ed. UFOs and the Alien Presence: Six Viewpoints. The 2020 Group, Visitors Investigation Project, 1991.

William Lyne. Space Aliens from the Pentagon: Flying Saucers Are Man-made Electrical Machines. Creatopia Productions, 1993.

Jim Marrs. Alien Agenda. HarperCollins, 1997.

Laton McCartney. Friends in High Places—The Bechtel Story: The Most Secret Corporation and How It Engineered the World. Simon & Schuster, 1988.

Robert D. McCracken. A History of Tonopah, Nevada. Nye County Press, 1990.

Jay Miller. Lockheed Martin Skunk Works: The Official History. Aerofax Inc., 1993. Revised edition, 1995.

Jay Miller. Lockheed's Skunk Works. The First Fifty Years—The Official History. Aerofax Inc., 1993.

Steve Pace. Lockheed Skunk Works. Motorbooks International, 1992.

Curtis Peebles. Dark Eagles: A History of Top Secret U.S. Aircraft Programs. Presidio Press, 1995.

Curtis Peebles. The Moby Dick Project. Smithsonian Institution Press, 1991.

Curtis Peebles. Watch the Skies! A Chronicle of the Flying Saucer Myth. Smithsonian Institution Press, 1994.

Chris Pocock. Dragon Lady: The History of the U-2 Spyplane. Motorbooks International, 1989.

Francis Gary Powers, with Curt Gentry. Operation Overflight. Holt Rinehart Winston, 1970.

Kevin D. Randle and Donald R. Schmitt. UFO Crash at Roswell. Avon, 1991.

Richard Rashke. Stormy Genius: The Life of Aviation's Maverick, Bill Lear. Houghton Mifflin, 1975.

Richard Rhodes. *Dark Sun*. Simon & Schuster, 1995.

Richard Rhodes. *The Making of the Atomic Bomb*. Simon & Schuster, 1986.

Ben R. Rich, with Leo Janos. *Skunk Works: A Personal Memoir of My Years at Lockheed*. Little, Brown and Company, 1994.

Jeffrey Richelson. *American Espionage and the Soviet Target*. William Morrow, 1987.

Jeffrey Richelson. *The U.S. Intelligence Community*. Ballinger, 1985.

Benson Saler, Charles A. Ziegler, and Charles B. Moore. *UFO Crash at Roswell: The Genesis of a Modern Myth*. Smithsonian Institution Press, 1997.

Richard Sauder. *Underground Bases and Tunnels: What Is the Government Trying to Hide?* Adventures Unlimited Press, 1995.

Alexander P. de Seversky. *Victory Through Air Power*. Simon & Schuster, 1942.

Michael Sherry. *The Rise of American Air Power: The Creation of Armageddon*. Yale University Press, 1987.

Edward A. Shils. *The Torment of Secrecy: The Background and Consequences of American Security Policies*. Ivan R. Dee, 1996.

Brian Shul and Walter Watson, Jr. *The Untouchables: Mission Accomplished*. Mach I Inc., 1991.

Erik Simonsen. *This Is Stealth: The F-117 and B-2—in Color*. Greenhill Books, 1992.

Michael Skinner and George Hall. *Red Flag: Air Combat for the 1990's*. Second edition. Motorbooks International, 1993.

Rebecca Solnit. *Savage Dreams*. Sierra Club Press, 1994.

John Steinbeck. *Bombs Away: The Story of a Bomber Team*. Viking, 1942.

Whitley Strieber. *Majestic*. Putnam's, 1989.

Bill Sweetman and James Goodall. *Lockheed F-117A: Operation and Development of the Stealth Fighter*. Motorbooks International, 1990.

Bill Sweetman. *Aurora: The Pentagon's Secret Hypersonic Spyplane*. Motorbooks International, 1993.

Bill Sweetman. *Stealth Aircraft: Secrets of Future Airpower*. Motorbooks International, 1986.

Bill Sweetman. *Stealth Bomber, Invisible Warplane, Black Budget*. Motorbooks International, 1989.

Evan Thomas. *The Very Best Men: Four Who Dared, the Early Years of the CIA*. Simon & Schuster, 1995.

Kenn Thomas, ed. *Popular Alienation: A Steamshovel Press Reader*. IllumiNet Press, 1995.

Keith Thompson. *Angels and Aliens: UFOs and the Mythic Imagination*. Addison Wesley, 1992.

A. Costandina Titus. *Bombs in the Backyard: Atomic Testing and American Politics*. University of Nevada Press, 1986.

U.S. Department of Energy. *Draft Environmental Impact Statement for the Nevada Test Site and Off-site Locations in the State of Nevada*. January 1996.

U.S. Department of the Air Force. *AFP 205-37. Preparing Security Classification Guides*. 1991.

U.S. Air Force, U.S. Bureau of Land Management. *Draft Environmental Impact Statement, Groom Mountain Range, Lincoln County, Nevada*. October 1985.

Renato Vesco and David Hatcher Childress. *Man Made UFOs 1944–1994: 50 Years of Suppression*. Adventures Unlimited Press, 1994.

Tim Weiner. *Blank Check*. Warner Books, 1991.

John F. Welch, ed. *RB-36 Days at Rapid City*. Silver Wings Aviation Inc., 1994.

Lt. George M. Wheeler. *Preliminary Report Concerning Explorations and Surveys, Principally in Nevada and Arizona*. U.S. Army Corps of Engineers, 1872.

Robert Wilcox. *Scream of Eagles*. John Wiley, 1990.

Robert Wilcox. *Wings of Fury*. Pocket Books, 1996.

Garry Wills. *Reagan's America: Innocents at Home*. Doubleday, 1985.

Fred Alan Wolf. *Parallel Universes*. Touchstone/Simon & Schuster, 1988.

Chuck Yeager, with Leo Janos. *Yeager*. Bantam, 1985.

PERIODICALS

Rick Atkinson, "Stealth: From 18-inch Model to $70 Billion Muddle: Project Senior C.J. The Story Behind the B-2 Bomber." *Washington Post*, Oct. 8, 1989, pp. A1–A38; Oct 9, 1989; pp. A1, A6–7; Oct 10, 1989, pp. A1, A14–15.

Andrew D. Basagio, "Area 51 and the CIA." *MUFON UFO Journal*. No. 291, July 1992, pp. 10–12.

Richard J. Boylan, "Secret 'Saucer' Sites." *MUFON UFO Journal*. No. 292, August 1992, pp. 14–15.

William J. Broad. "Wreckage in the Desert Was Odd but Not Alien." *New York Times*, Sept. 18, 1994, pp. 1, 40.

Stuart Brown. "Searching for the Secrets of Groom Lake." *Popular Science*, March 1994, pp. 52–54, 84–85.

Stuart Brown and Steve Douglass. "Swing Wing Stealth Attack Plane." *Popular Science*, January 1995, pp. 54–56, 86.

Malcolm W. Browne. "Rumors of U.S. Superplane Appear Unfounded." *New York Times*, Jan. 19, 1993, p. C8.

Jane Castro. "Grapevine." *Time*, May 25, 1992.

Jon Christenson. "How Military Secrecy Zones Out Nevada." *High Country News*, Dec. 27, 1993.

John Connolly. "Inside the Shadow CIA." *Spy*, September 1992, pp. 46–54.

Elise DeMan. "Shooting Stealth." *Air & Space*, pp. 92–94.

Michael Dornheim. "United 747 Crew Reports Near-Collision with Mysterious Supersonic Aircraft." *Aviation Week and Space Technology*, Aug. 24, 1992, p. 24.

Steve Douglass. "Flying Artichoke." *Popular Science*, December 1994, p. 16.

John J. Fialka. "Clinton to Disclose Tab for Spying, Propose Overhaul." *Wall Street Journal*, April 24, 1996, p. 1.

David A. Fulghum. "Groom Lake Tests Target Stealth." *Aviation Week and Space Technology*, Feb. 5, 1996, pp. 26–27.

David Fulghum. "Payload, Not Airframe Drives UCAV Research Oversight." *Aviation Week and Space Technology*, June 2, 1997, pp. 51–53.

Marian Green. "Unions Win Representation Elections for Workers at Groom Lake." *Las Vegas Review Journal and Las Vegas Sun*, Feb. 17, 1996.

Gerald Haines. "A Die Hard Issue: CIA's Role in the Study of UFOs, 1947–90." *Studies in Intelligence*. Central Intelligence Agency, Langley, Virginia.

Roy J. Harris, Jr. "Evidence Points to Secret U.S. Spyplane." *Wall Street Journal*, Dec. 4, 1992, p. B6.

Peter Hellman. "The Little Airplane That Could." *Discover*, February 1987, pp. 78–87.

Fred L. Humphrey. "Geology of the Groom District Lincoln County Nevada." Nevada State Bureau of Mines. *University of Nevada Bulletin*. June 1945, vol. XXXIX, no. 5.

Margaret A. Jacobs. "Secret Air Base Broke Hazardous-Waste Act, Workers' Suit Alleges." *Wall Street Journal*, Feb. 8, 1996, p. 1.

Steve Kanigher. "Area 51 Saga Heads to Federal Court." *Las Vegas Sun*, Nov. 3, 1997.

Frank Kuznik. "Aliens in the Basement." *Air & Space*, August/September 1992, pp. 34–39.

John Lear. "The Grand Deception: How the Gray EBE's Tricked MJ-12 Into an Agreement." *CUFORN Bulletin*, March–April 1989.

Samuel W. Matthews. "Nevada Learns to Live with the Atom." *National Geographic*, August 1953, pp. 839–50.

Thomas P. McIninch (pseud.). "The Oxcart Story." *Studies in Intelligence*, Winter 1970–71, Central Intelligence Agency, Langley, Virginia.

John T. McQuiston. "Plot Against L. I. Leaders Is Tied to Fear of U.F.O's." *New York Times*, June 22, 1996.

Peter Merlin. "Dreamland—The Air Force's Remote Test Site." *Aerotech News*, April 1, 1994.

Peter Merlin. "Secret Base in Nevada Desert Suffered Effects of Nearby Nuclear Testing." *Aerotech News*, Oct. 20, 1995.

Peter Merlin. "Test and Decontamination Revisited: Operation Plumbob and Project 57." *Aerotech News*, Dec. 15, 1995.

Gary Paine. "A Mine, the Military and a Dry Lake: National Security and the Groom District." *Lincoln County, Nevada Historical Society Quarterly*. Vol. 39, no. 1, pp. 20–42.

Phil Patton. "Exposing the Black Budget." *Wired*, November 1995, pp. 94–102.

Phil Patton. "Robots with the Right Stuff." *Wired*, March 1996, pp. 148–51, 210–15.

Phil Patton. "Stealthwatchers." *Wired*, February 1994, pp. 78–83, and "A Visit to Dreamland," pp. 80–81.

Gregory Pope. "America's New Secret Aircraft." *Popular Mechanics*, December 1991, p. 34.

Eileen White Read. "They Sneak Around, Learning What They Can About Stealth." *Wall Street Journal*, April 26, 1988, pp. 1, 25.

Randall Rothenberg. "Area 51, Where Are You?" *Esquire*, September 1996, pp. 88–97.

Rhonda L. Rundle. "Lockheed Employees Health Complaints Prompt Inquiries by 2 Federal Agencies." *Wall Street Journal*, Oct. 3, 1988, p. 1.

Murray Schumach. " 'Disk' Near Bomb Test Site Is Just a Weather Balloon." *New York Times*, July 9, 1947, pp. 1, 10.

Murray Schumach. "Disks Soar over New York, Now Seen Aloft in All Colors." *New York Times*, July 8, 1947, pp. 1, 46.

William B. Scott. "Black Projects Must Balance Cost, Time Savings with Public Oversight." *Aviation Week and Space Technology*, Dec. 18, 1989, pp. 42–43.

William B. Scott. "New Evidence Bolsters Reports of Secret, High-Speed Aircraft." *Aviation Week and Space Technology*, May 11, 1992, p. 62.

William B. Scott. "Recent Sightings of XB-70–like Aircraft Reinforce 1990 Reports from Edwards Area." *Aviation Week and Space Technology*, Aug. 24, 1992, pp. 23–24.

William B. Scott. "Secret Aircraft Encompasses Qualities of High-Speed Launcher for Spacecraft." *Aviation Week and Space Technology*, Aug. 24, 1992, p. 25.

William B. Scott. "Spooks—It's Time for a Revelation." *Aviation Week and Space Technology*, Dec. 22/29, 1997, p. 96.

William B. Scott. "Triangular Recon Aircraft May Be Supporting F-117A." *Aviation Week and Space Technology*, June 10, 1991, p. 20.

Bill Sweetman, "The Invisible Men." *Air and Space*, May 1997, pp. 19–27.

John Tierney. "The Real Stuff." *Science 85*, September 1985, pp. 24–35.

Donovan Webster. "Area 51." *New York Times Magazine*, June 26, 1994, p. 32.

Christopher Weir. "Paint It Black." *Metro*, Jan. 9–15, 1997.

James A. Williams. "Scanning the USAF's 'Area 51' Mystery Base." *Popular Communications*, April 1995, pp. 8–9.

"Dumb Pursuit of a Smart Weapon." *New York Times*, March 28, 1988, p. 18.

"Eisenhower Plays Golf on Vacation." *New York Times*, Feb. 20, 1954.

"The Mystery at Groom Lake." *Newsweek*, Nov. 1, 1993, p. 4.

"Out of the Clouds: Secret Stealth Fighter Gets More Exposure." *Wall Street Journal*, Dec. 27, 1989, p. 1.

"Possible 'Black' Aircraft Seen Flying in Formation With F-117s, KC-135s." *Aviation Week and Space Technology*, March 9, 1992, pp. 66–67.

"Scientists' and Engineers' Dreams Taking to Skies as 'Black' Aircraft." *Aviation Week and Space Technology*, Dec. 24, 1990.

"Secret Advanced Vehicles Demonstrate Technologies for Future Military Use." *Aviation Week and Space Technology*, Oct. 1, 1990, p. 20.

"Skunk Works Revenues Point to Active Aurora Program, Kemper Says." *Aerospace Daily*, July 17, 1992, p. 102.

"Tesla at 78 Bares New 'Death-Beam.' " *New York Times*, July 11, 1934.

"TR-3A Evolved from Classified Prototypes, Based on Tactical Penetrator Concept." *Aviation Week and Space Technology*, June 10, 1991, p. 20.

WORLD WIDE WEB SITES

Keeping up with new discoveries in Dreamland is best done through sites on the World Wide Web. The following are the most useful, and all contain links to other sites of interest:

Glenn Campbell's exhaustive and indispensable Area 51 and UFO site: http://www.ufomind.com/area51/

Steve Douglass's Project Black site on black aircraft, plus the *Intercepts* newsletter:
     http://www.perseids.com/projectblack/
The Federation of American Scientists: http://www.fas.org/
Dan Zinngrabe's site with extensive histories of Aurora, the TR3A, and Tier III:
     http://www.macconnect.com/~quellish
Andreas Gehrs-Pahl's extensive aviation site:
     http://www.umcc.umich.edu/~schnars/aero.htm
Paul McGinnis ("Trader") and the Freedom Ridge Oversight Council:
      http://www.frogi.org/
UFO Folklore: http://www.qtm.net/~geibdan
Parascope: http://www.parascope.com/
Tom Mahood's Blue Fire page: http://www.serve.com/mahood/bluefire.htm/

Index

About the Author

PHIL PATTON is the author of *Made in USA: The Secret Histories of the Things That Made America*, selected by the Book-of-the-Month Club; *Open Road*, named a *New York Times* Notable Book of the Year; *Voyager* (with Jeana Yeager and Dick Rutan); and other books.

He writes the "Public Eye" column for *The New York Times* and is a contributing editor of *Esquire*, *Wired*, and ID magazines.

Mr. Patton has taught at the Columbia Graduate School of Journalism and participated in a number of public television series. He grew up in North Carolina and was educated at Harvard and Columbia. He has also developed World Wide Web pages for *Esquire*, the New York Web, and Hearst New Media.

Visit Phil Patton at www.philpatton.com.